Springer Praxis Books

Popular Science

The Springer Praxis Popular Science series contains fascinating stories from around the world and across many different disciplines. The titles in this series are written with the educated lay reader in mind, approaching nitty-gritty science in an engaging, yet digestible way. Authored by active scholars, researchers, and industry professionals, the books herein offer far-ranging and unique perspectives, exploring realms as distant as Antarctica or as abstract as consciousness itself, as modern as the Information Age or as old our planet Earth. The books are illustrative in their approach and feature essential mathematics only where necessary. They are a perfect read for those with a curious mind who wish to expand their understanding of the vast world of science.

More information about this subseries at http://www.springer.com/series/8158

Alex Ely Kossovsky

The Birth of Science

 Springer

Published in association with
Praxis Publishing
Chichester, UK

Alex Ely Kossovsky
New York, NY, USA

Springer Praxis Books
ISSN 2626-6113   ISSN 2626-6121   (electronic)
Popular Science
ISBN 978-3-030-51743-4   ISBN 978-3-030-51744-1   (eBook)
https://doi.org/10.1007/978-3-030-51744-1

This Springer imprint is published by the registered company Springer Nature Switzerland AG
The registered company address is: Gewerbestrasse 11, 6330 Cham, Switzerland

Kepler's Celestial Data Analysis,
Galileo's Terrestrial Experiments,
Newton's Grand Synthesis,
Midwifing the Birth of Science.

If I have seen further than others, it is by standing upon the shoulders of giants.
Isaac Newton

The senses deceive from time to time, and it is prudent never to trust wholly those who have deceived us even once.
René Descartes

If you would be a real seeker after truth, it is necessary that at least once in your life you doubt, as far as possible, all things.
René Descartes

Geometry was co-eternal with the Divine Mind prior to the birth of all things.
Johannes Kepler

Mathematics is the language in which God wrote the universe.
Galileo Galilei

Preface

This book has been written for two distinct audiences, and without the slightest internal contradiction, namely, for the educated general public who are not necessarily proficient in mathematics and physics, as well as for expert physicists, other scientists, mathematicians, and statisticians, who I hope will find many sections of the book to be thought-provoking and informative at their own professional level. The four chapters marked as 'optional' require a somewhat higher level of mathematical skills on the part of the reader, and could be skipped by the general reader without any loss of continuity, and without diminishing their ability to understand the rest of the book. It should be noted that the notation for the multiplication of A by B is conveniently denoted here in a variety of ways, such as AxB, A*B, AB; and that conventionally, squares, namely NxN, are denoted as N^2, and cubes, namely NxNxN, are denoted as N^3. Feedback, questions, suggestions, or criticism about the content of the book could be sent to the email address akossovsky@gmail.com, and would be warmly welcome by the author.

New York, USA   Alex Ely Kossovsky

Acknowledgment

I would like to express my profound gratitude to the late Professor Edward Binkowski, my mentor and the former director of the Applied Mathematics and Statistics Department at the City University of New York, who passed away last year. I thank him for teaching a broader, more balanced and open-minded approach regarding issues in mathematical statistics and philosophy of science, as well as for his support and encouragement leading eventually to the writing of this book. I would also like to extend my heartfelt thanks to Professor George Feuerlicht of Prague University of Economics and Sydney University of Technology, for his unfailing loyalty, support, and enthusiasm for this book, as well as for his thorough and brutal purge of typos from my original manuscript. Lastly, I would like to express my gratitude and appreciation to the distinguished physicist and author Professor Don Lemons, and to Emeritus Professor of Cognitive Sciences Walter Makous, for their scientific insights and comments, for their constant support, and for Walter's reviewing of the prose in the book.

Contents

Chapter 1
The Scholars Who Sequentially Ignited the Scientific Revolution

It took humanity two centuries, five intellectual giants, and one meticulous astronomer who carefully recorded the movements of the planets across the sky, to discover the fundamental laws of motion known as modern physics. During that multigenerational process, the scientific method was adopted along the way, and mathematics was endowed a central role in science. The six significant personalities in this drama were:

Nicolaus Copernicus	(1473–1543)
Tycho Brahe	(1546–1601)
Johannes Kepler	(1571–1630)
Galilei Galileo	(1564–1642)
René Descartes	(1596–1650)
Isaac Newton	(1642–1727)

It was the continent of Europe, during the transformative era of the Renaissance, where this remarkable revolution in knowledge took place. Yet, Europe at that era was neither peaceful nor prosperous; rather it was largely a continent suffering from superstition and ignorance; ruled mostly by exploitative and despotic kings. More benevolent and humanistic rulers who supported the arts and the sciences were the exception. It was also a continent torn by brutal and senseless wars, mostly in order to enhance this or that dynastic royal family, and also in part due to the violent schism between the newly formed Lutheran Church established by Martin Luther in the Protestant Reformation, and the older and well-established Catholic Church.

Martin Luther famously nailed his Ninety-Five Theses calling for the reform of the Church in 1517 to the door of the cathedral in Wittenberg, Germany, an act that could have easily cost Luther his life. Luther's Reformation movement against the Church was a response to the occurrences of abuse and corruption on the part of many in the clergy, demanding payments for forgiveness of sins to ensure eternal happiness in the afterlife. Luther's Protestant movement spread like wildfire in Europe, and it led to prolonged and numerous wars.

© The Editor(s) (if applicable) and The Author(s), under exclusive license to Springer Nature Switzerland AG 2020
A. E. Kossovsky, *The Birth of Science*, Springer Praxis Books,
https://doi.org/10.1007/978-3-030-51744-1_1

Most notable was the Thirty Years' War which was fought mainly in Central Europe between 1618 and 1648, involving most of the European great powers. It was one of the longest and most destructive conflicts in all of human history resulting in eight million European fatalities.

In addition to wars, the Great Bubonic Plague, better known as the Black Death, was continuously hanging menacingly over the continent with frequent recurrences, resulting in horrific death toll. It was the most devastating pandemic recorded in human history. The European population was reduced by between one-third and two-thirds. The plague created religious, social, and economic upheavals, with profound effects on the course of European history, and it might have played a positive albeit indirect role in the emergence of the Renaissance.

Superstition and cruelty reigned supreme all over this charming continent where horrific witch-burning occurred frequently. The monetary incentives that the accusers and the Inquisition clergy gained from each case, confiscating the entire property of the accused 'witch', further promoted and increased the frequency of these evil occurrences.

Yet, in spite of these political, religious, and pandemic oppressions, the Europeans also displayed their rebellious intellectual spirit, their tendencies for independent and individualistic way of thinking and doing things, and this strongly facilitated the emergence of the Scientific Revolution.

Let us now briefly narrate the sequential contributions to the Scientific Revolution of the six remarkable personalities mentioned above.

Copernicus' formulation of the simple and straightforward heliocentric model in astronomy—where Earth and the planets revolve around the Sun—produced the essential and workable foundation upon which Kepler was able to formulate his three laws of planetary motion. This simple celestial model also hinted at the gravitational pull exerted on the planets by the Sun, although initially this was not so obvious.

The heliocentric model was passionately supported by Galileo, who risked his life defending it. Galileo then gave the model decisive confirmations with his telescopic astronomical observations, especially with his discoveries of Jupiter's four moons and the phases of Venus, which are essentially the same type of phenomenon as the Moon's phases. Thereafter, the geocentric model of astronomy summarized in the ancient treatise 'Almagest' by Ptolemy of Alexandria, where the planets and the Sun revolve around the Earth, began to be discredited and finally abandoned.

Galileo's telescopic observations of the moon described its surface as being uneven and irregular, crowded with depressions and bulges, hills and valleys, and that overall its features appeared quite earth-like.

If the moon is such, then the planets might have earth-like features as well, and perhaps Earth itself could be just another moving planet! Perhaps Earth is not special at all in astronomical sense.

Galileo also discovered the existence of sunspots. They first appear quite small on the surface of the Sun, and then they expand, and finally they fade away completely—repeating such cycles in an irregular way. His observations showed that these were not clouds or small moons but actual features of the surface of the Sun itself, and that the Sun rotates—which is consistent with the observation of the continuous movements of these spots on its surface.

All this contradicted the ancient Greek philosophy and Aristotelian claim about the perfection and immutability of the heavens and all celestial bodies, setting them apart from the imperfect, changeable, and lowly Earth.

Another false belief of the ancient Greek culture that Galileo dismantled was the generally accepted Aristotelian hypothesis that a moving body needs continuous force or some kind of an agent of change in order to keep it in motion. In contrast, Galileo developed the concept of 'inertia', namely that if an object is freely moving horizontally by itself, it will continue to do so unless something acts upon it to stop it, to accelerate it, or to change its direction. To us in the modern world, the concept of inertia seems more acceptable and even quite natural. We have ice-skating rinks, rockets, missiles, airplanes, and especially communication satellites, all of which promote the idea of inertia. In addition, our ability to observe almost directly so much of the astronomical universe with telescopes, satellites, or space stations, reinforces the idea that motion is quite natural in the universe, and that every star, planet, galaxy, asteroid, or meteoroid, always moves, rotates, and orbits. In contrast, for the earth-bound ancients and even for the non-modern civilizations, motion always seemed to stop quickly and to involve a great deal of effort. Everything seemed sluggish and full of friction. Even snow balls which rolled downhill quickly and effortlessly, soon stop at the bottom of the hill; and tree leaves which easily fell down, always stop and rested on the ground. They invented the wheel, to ease motion, yet they still needed to labor hard in pushing those carts and carriages, or else they needed to find a horse to pull the heavy wagon.

In addition to all that, Galileo's insistence on the scientific method, emphasizing observations and experiments as opposed to pure meditation and abstract philosophical reasoning about the physical world, was a giant step forward. Galileo's additional work on the acceleration of falling bodies and projectile motion was also decisive for Newton's later discoveries.

Tycho Brahe's careful astronomical observations which were by far more accurate than the best available observations at the time, gave Kepler the raw material for his planetary analysis, namely the appropriate data, upon which to build his work successfully.

Kepler's 1st law of planetary motion conceives of elliptical orbits instead of circular ones and positions the Sun slightly away from the center of these elliptical orbits, namely at one of the two focal points. Kepler's 2nd law of planetary motion states that each planet speeds up while its orbit is near the Sun and slows down while its orbit is farther away. Kepler's 3rd law of planetary motion relates to the totality of the data set of the six known planets of that era and leads to the statistical discovery that for each planet, the square of the time for one full orbit (T^2) is equal to the cube of its distance from the Sun (D^3). This is expressed algebraically as $T^2 = K \times D^3$, where dimension constant K is necessary to adjust for whichever types of units the astronomer might be using. Speed, which is defined as distance travelled per unit of time ($S = D/T$), for example kilometers per hour, is then easily calculated for each planet, and this points to positive correlation between the planet's speed and its closeness to the Sun. This implies that the inner planets Mercury and Venus which are very near the Sun orbit much faster than the outer planets Jupiter and Saturn which are much farther away. Hence, Kepler's 3rd law which elegantly and precisely accounts for the variation in planetary speed in terms of distance from the Sun, provided yet an even stronger hint at gravitational interactions between the Sun and the planets. Mercury and Venus are closer to the Sun and thus the effect of gravity is more pronounced, and this perhaps could explain why they move fast, while Jupiter and Saturn are farther away from the Sun and thus the effect of gravity is less pronounced, and this perhaps could explain why they move slower.

René Descartes appealed to rationality, and called for an open-minded approach to science, insisting on never accepting anything as true unless clearly demonstrated to be such. Descartes urged healthy skepticism and warned about intellectual prejudice and generational-inertia where false beliefs and misguided ideas are inherited from one generation to the next generation. Hence Descartes considerably improved the philosophical background and approach in scientific research in general. Descartes' invention of the Cartesian Plane—combining the vertical and the horizontal dimensions to create the graph or the chart—provided the foundation upon which calculus was developed by Newton. Descartes' modern treatment of independent variables also contributed to Newton's mathematics.

It was Newton who put it all together with his three laws of motion and the law of universal gravitation; and it was Newton who in effect gave birth to modern physics we call classical mechanics, which is so successfully and practically being applied here on Earth in all our industrial activities. The Industrial Revolution which started around 1760 in England, and which then spread throughout Europe and the United States, was the direct consequence of the intellectual efforts of these six personalities and the birth of science.

Chapter 2
Was It an Apple, Moon, or Planets for Newton?

A typical star has on average only one to two accompanying planets, and current astronomical opinion is that approximately only 1 in 10 stars are blessed with multiple planets such as our Solar System which possesses at least 8 planets. Apparently, the abundance of planets around our star (the Sun) is a bit rare in the cosmos. Such planetary abundance led to many inquiries and careful observations, and prodded our scientific discoveries.

The moon and modern communication satellites circling the Earth, as well as the planets circling the Sun, all move forward due to inertia, but are also being pulled inward due to gravity. Such a tug of war between two opposing tendencies could be perfectly balanced under the right speed resulting in perpetual orbit where neither inertia nor gravity has the upper hand.

Figure 2.1 illustrates the three distinct outcomes (depending on the initial launch speed) that could possibly occur when a cannon ball is fired from a very high mountain or tall platform. Figure 2.2 depicts the Moon circulating the Earth under the equal and balanced influences of the gravitational force and the inertial tendency, maintaining its perpetual motion without any need for an engine, fuel, or any other energy source.

Let us then imagine humanity here on Earth without any other planets and without a moon either, and let us envision the indirect impact of the absence of these heavenly bodies on the pace of scientific awakening (but not on its eventual emergence).

The wandering planets in the sky against the steady backdrop of the stars left the ancients puzzled and prodded them to theorize about the phenomenon and carefully measure their movements.

It is almost a certainty that the correct heliocentric view owes to these planets its relatively early emergence around the Hellenistic Period (Aristarchus and Philolaus) and more decisively so during the Renaissance (Copernicus, Kepler, and Galileo). It would certainly have taken us longer to figure out that Earth—had it been a single planet—is circling its star annually rather than the other way around. Kepler's whole lifework was exclusively about the planets. His three planetary laws, though merely descriptive and statistical in nature, devoid of any universal theory, strengthened the heliocentric view considerably at the time and laid the foundation for subsequent

A. E. Kossovsky, *The Birth of Science*, Springer Praxis Books, https://doi.org/10.1007/978-3-030-51744-1_2

Fig. 2.1 Three possible outcomes according to speed

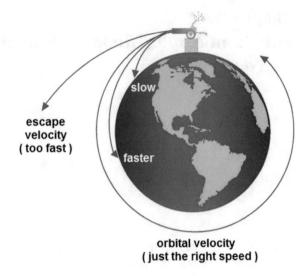

Fig. 2.2 Perfectly balanced tug of war—gravity and inertia

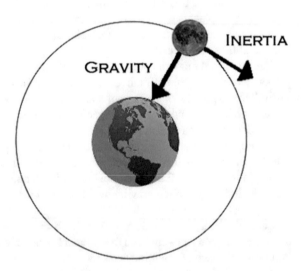

discoveries. Without the planets, Kepler's laws would have never been proclaimed, Newton might not have had his grand insight, and humanity's progress in science would have been slower.

Indeed, the planets played a direct role in the discovery of Newton's mechanics. Galileo was already talking about inertia and hinting about forces, but had encountered difficulties. "If only there were no friction and no resistance, just for a short while, or just for some particularly polished, smooth, and very fluid device, then everyone from Pisa to Rome would fully observe my principle of inertia for horizontal motion" Galileo kept promising everybody. But Galileo's misfortune was that he had to work with frictional rolling slopes and pendulums that slow down and quickly stop moving. His adversaries would easily contradict his principle of inertia with a simple demonstration of kicking a rock forward on the rough and bumpy ground, and observe it stopping soon afterwards. Even his work and observations with the smoother and almost frictionless projectile motion such as cannon balls and forwardly thrown objects in the air was problematic and difficult, since such motion only lasted several seconds up in the air or a few minutes at most, and was exceedingly difficult to measure with the primitive instruments of that era. In addition, Galileo was at a complete loss with regard to vertical motion, which always accelerated downward and which appeared to totally contradict his general idea of inertia. Horizontally, things always slowed down and quickly stopped, while vertically things moved quite energetically, and they were even gaining speed (until hitting the ground).

Galileo's three greatest terrestrial discoveries were about horizontal inertia; about vertical gravitational free fall; and about projectile motion. Regarding vertical free fall, Galileo discovered that the acceleration is a constant and independent of the weight of the falling object, so that light bodies fall just as fast as do heavy bodies. True, light feathers or paper fall down quite slowly due to air resistance, while heavy metal balls and rocks accelerate downward very fast. Ignoring the tiny effect of air resistance, a rock weighing 5 kg should reach the ground at the same time as a heavier rock weighing 30 kg falling from the same height. Without air or any other resistance, all bodies should fall down at the same rate and speed. Regarding measurements and analysis of projectile motion, Galileo discovered that it should be split into its vertical and horizontal components, and that indeed the forward horizontal component of motion was measured to be approximately constant and inertia-like, while the vertical component of motion was measured to be approximately with the same rate of acceleration as for freely falling bodies such as rocks dropped straight downward from the roof of a big tower. Galileo's detractors would criticize him by pointing to this profound dichotomy in his entire work, and demand either absolute inertia for both dimensions, for vertical as well as for horizontal motion; or constant acceleration in both dimensions, consistently. His worst enemies would taunt him by proposing an inverted model—that of vertical inertia together with horizontal acceleration (but to which direction? North? East?). Galileo was unable to defend his work within such grand (unifying) philosophical argument, but he firmly stood by his observations and his analysis.

Around the year 1638, all of Galileo's work was already accomplished, and there would be no progress in the field until the year 1665 when young Newton formulated his three laws of motion and universal gravitation. Galileo, the first true physicist, has revolutionized philosophy of science in general, for all disciplines, with his emphasis on experiments as opposed to pure meditations, as well as his adoption of mathematics as an essential tool in science. He has spent a lot of his energies on the celestial world; and was the first human to gaze at the Moon, the Sun, the planets, and the planets' satellites via a telescope. Yet, none of his celestial discoveries were helpful in directly deducing the theory of motion itself, the core concern of physics. Galileo's contributions to physics, his analysis of motion, were all firmly earthbound, relying exclusively on his terrestrial experiments and measurements, and it wasn't complete. Science needed another giant, one who would turn also to the heavens in order to facilitate the final push towards a complete theory of motion. We therefore note with amazement that the year of Galileo's death, 1642, was also the year of Newton's birth. Galileo on his last day, blind from telescopic observations of the Sun, still surrounded by Inquisition guards outside his house, passed away on 8 January 1642. Later that year, on 25 December 1642, Isaac Newton was born, prematurely and thus exceedingly small and vulnerable, but he survived, and lived to continue Galileo's work and make his grand discoveries.

It was at this crucial crossroad that the planets were eagerly awaiting to play their indispensable prodding role, **being the only friction-less and smooth system or phenomenon we humans could have observed back then, and possessing near perpetual motion, a fine manifestation of the principles of inertia and gravity combined**. Here comes Newton, integrating and harmonizing the celestial with the terrestrial, inspired in particular by the images of planets swinging ceaselessly around the Sun, noting Kepler's planetary observations and especially how speed correlates with closeness to the Sun, noting Galileo's concepts and observations of motion on Earth, and especially of equal downward vertical acceleration regardless of weight, versus horizontal inertia, inventing gravity, putting it all together and formulating classical mechanics.

By insisting on universal inertia in all directions and in all dimensions, vertical, horizontal, or even diagonal, Newton was able to present a wholesome and unified theory, while allowing for exceptions to the principle of inertia in the case where forces were present, such as gravitational forces, or Electrical and Magnetic forces which would be discovered in the following nineteenth century after his death, as well as the Weak and Strong nuclear interaction forces inside the atom to be discovered in the twentieth century. For Newton, the gravitational force just happened to be 'vertical' on the local flat surface of the Earth simply because the force of gravity points down to the center of the spherical Earth.

Newton's profound insight in stating (Force) = (Mass) × (Acceleration) in terms of any generic force, and not only of the gravitational type, allows the law to be universally applied for all the 4 known forces. In addition, by stating his law in a vector-directional form, Newton is actually stating 3 distinct $F = M \times A$ laws, one for each of our 3 spatial dimensions—and without a doubt this particular insight of Newton was directly inherited and inspired by Galileo's successful split of projectile motion into 2 distinct dimensions—the horizontal and the vertical.

Earth as a lone planet together with a single moon would not have been as beneficial as the planets, since then Kepler would have nothing to chew on, and possessing merely a moon may not have been enough to inspire Newton. Although, a moon would certainly offer some hint if we had no planets at all to guide us. Mere terrestrial apples falling on thinking heads certainly weren't enough of an inspiration for such a great feat, the legend notwithstanding.

If the legend of the apple falling on Newton's head was indeed true, most likely he was not sitting directly beneath the tree, but rather several feet away, and the apple was not dropping, but rather flying sideway after being kicked with a pole—manifesting for Newton (terrestrial) projectile motion which neatly combines gravity and inertia (under the gravitational influence of the Earth), and perhaps prompting him to realize that planetary motion could also be thought of as (celestial) projectile motion (under the gravitational influence of the Sun).

The planets also had a second prodding role, **being the only phenomenon with clearly differentiated forces and differentiated resultant accelerations we humans could have observed back then**. Galileo's work on free fall terrestrial gravity was only of the fixed gravitational acceleration rate experienced on the surface of the Earth which is 9.807 m/s^2. Galileo could not have experienced variations in gravitational acceleration, even by experimenting on top of Monte Bianco in the Alps, the highest mountain in Italy, rising almost 5 km above sea level. This is so because the distance to the center of the Earth there in the Alps is only a tiny bit greater than the distance to the center of the Earth at sea level which is 6371 km, and such minuscule decrease in gravitational acceleration on top of the Alps could not have been measured even if attempted. In addition, Galileo did not refer to any gravitational 'forces', only to gravitational acceleration. Without referring to the motions of the planets, and only experimenting with gravity on Earth, Galileo's work could never hint at variations in the gravitational force due to distance, namely the inverse square law in the relationship between the gravitational force and the distance between the two attracting objects. It was Kepler's work on the celestial planets that gave the vague hint of forces per se; and of differentiated types of forces in particular. The six planets known at that era possess a great variety of distinct time periods of revolution, distinct distances from the Sun, and district speeds. Taken together they strongly hint that the force of gravity is real and that it diminishes with distance from the Sun.

It seems that before Kepler gave his concise summary of the relationship between orbital distances and time periods, no one bothered to note that the inner planets Mercury and Venus move quite fast, and that the outer planets Jupiter and Saturn move much more slowly, even though it should have been obvious even before Kepler, and could have been easily calculated. Perhaps the lengthy time period of the outer planets was explained away or excused solely in terms of the need to traverse wider and longer circumferences, and not in terms of possessing lower speeds as well (which reinforces the 'delay' in completing one full revolution). Yet nobody bothered to sit down and calculate this, or any speed for that matter, in order to attempt to verify even this erroneous idea of uniform speed for all the planets. Most likely nobody even thought of the very concept of 'planetary speed', which might have seemed as an alien idea back then. The ancients (and we) never observed directly any planet that is physically 'moving', only that their positions in the sky are changing each day or each week, and thus logically they must be 'moving' up there in the sky, but all this was perhaps a bit too abstract for them. It seems that Kepler's second and third laws simply drove the point home to those with acute minds who paid strong attention (like Newton), namely that the planets possess differentiated speeds that strongly and positively correlate with closeness to the Sun.

How intriguing it is to note—through the conceptual lens of such a historical vista—that it took watching the skies for ages, painstakingly recording those five faint little points of light, far away from us, visible only at night when all are sleepy and tired, and only when the Moon and the clouds are not there to hinder observations, to get a hint about mechanics, mechanics that always worked fine here on earth, right in front of our inept eyes! We patiently watched and recorded those far away planets for many generations, so that finally we would be able to profitably operate machines and industry here on Earth, years and generations later.

People like Tycho Brahe, and all before him, who spent years observing and recording the planets, could not have imagined the unexpected discoveries and the deep understanding of nature in general that their work would lead to. They did their work sensing that it was important and worthwhile to study those unreachable far away heavenly dim points of lights, even though at the time it might have seemed to be such an irrelevant and futile exercise since we couldn't fly or even jump more than a foot or two up in the air. Surely many astronomers of the ancient eras were ridiculed and belittled as totally impractical and useless eccentrics, spending all their time and efforts on such irrelevant and silly endeavor. Yet, their work proved highly relevant and immensely practical!

The planets' only adverse contribution to confusion and superstition was the rise they gave to the 'study' of astrology.

But then again, we humans were always keen on twisting facts as well as reason itself, and would have been using other less intricate natural timing systems to predict future events by erroneously correlating such natural occurrences with past human events of major importance.

Chapter 3
Galileo's Other Revolution

Galileo's conflict with the church of his era was a drama which played a decisive part in that fateful event of modern history, namely the secularization and rationalization of thought. In a sense, besides being the scientist, Galileo was also the humanist, trying to bring on awareness of the new scientific ideas, only to be forcefully opposed by the religious fundamentalism of his era. That drama is perhaps best told by the late historian Giorgio de Santillana in his fascinating book 'The Crime of Galileo'.

The objection of the church to the heliocentric view seems to us nowadays in retrospect senseless, as if the church was shooting itself in the foot, stubbornly fighting a losing cause. What possible harm to religion or morality was there in us revolving around the Sun, or in finding hills and valleys on the Moon?

The official and purported account of the church objection was based on very few and quite minor and insignificant passages in the Bible such as:

> Then Joshua spoke to the Lord in the day when the Lord delivered up the Amorites before the children of Israel, and he said in front of Israel: Sun, stand still over Gibeon! and Moon in the Valley of Ayalon! And the Sun stood still, and the Moon stopped, till the people had revenge upon their enemies. (Joshua 10:13)

> The words of the Teacher, son of David, king in Jerusalem. Vanity of vanities, said the Teacher; vanity of vanities, all is vanity! What do people gain from all their labors at which they toil under the Sun? One generation passes away, and another generation comes, but the Earth remains forever. The Sun rises and the Sun sets, and to its place it tends again, it rises there. (Ecclesiastes 1:5)

> The Lord set the Earth on its foundations; it can never be moved. (Psalm 104:5)

Surely the church could have assigned these passages metaphorical and figurative meanings, instead of taking them literally. Yet, these supposedly offended passages mask the real reason for the church objection. Rather, the objection surely had been based on much deeper grounds. First, it goes back to the Greeks and Romans who elevated and divined celestial objects, so much so that 'up' came to mean holy, light, perpetual and pure; while 'down' came to mean heavy, sinful and transient. People

© The Editor(s) (if applicable) and The Author(s), under exclusive license to Springer Nature Switzerland AG 2020
A. E. Kossovsky, *The Birth of Science*, Springer Praxis Books,
https://doi.org/10.1007/978-3-030-51744-1_3

would refer to God by pointing upward toward the sky, in spite of the fact that the earth is round and that it rotates.

Aristotle certainly has made matters worse with his pseudo theories deeming the Moon, the Sun, and all celestial bodies to be perfect, pristine, and spotless, and to move about along perfect circular paths. Since antiquity celestial immutability was a fundamental axiom of the Aristotelian world-view, postulating that the heavens were eternally unchangeable. Any offending and disrespectful declaration or remark against a celestial object was comparable to heresy and the desecration of the divine.

Aristotle's view about the phases of the Moon and its eclipses, that "it's in the nature of the Moon" (which explains nothing!) fits very well with his other doctrines, since merely trying to explain and analyze the Moon reduces it to something less important. To be divine means to have no explanation and to be beyond any analysis, to be an enigma. To an Aristotelian, the very notion of "physics of the sky" was contradictory and anathema, and comparable to desecrating the divine and holy. Only tainted and lowly terrestrial objects are allowed to come under the scrutinizing and severe gaze of the Greek philosopher for analysis.

The second important factor was the basic Judeo-Christian tenet that the whole creation was about us, centered on human conduct and choices, and therefore the Sun, the stars, the planets, and the Moon, were all created for the inhabitants of the Earth, and so it was natural to think of them as physically revolving around us, around important and special Earth and its significant inhabitants, and not vice versa.

Did anyone notice the subtle conceptual contradiction? Celestial bodies were divine and perfect yet obediently and submissively are revolving around us, serving us, in spite of their great divinity!

Enters Galileo, pointing to blemishes on the Moon to discredit her purity; finding transient, changeable, and irregular spots on the Sun to ruin its reputation as an immutable and perfect body; discovering rebellious and offending moons of Jupiter that do not revolve around us, but rather around Jupiter; and then delivering the final harsh blow, making us totally unimportant by moving and revolving around the Sun, thereby threatening the whole social and religious order of the time. We might wonder then how Galileo managed to escape with his life, as he only got house arrest, condemned to be surrounded by Inquisition soldiers for the rest of his life.

Chapter 4
Ancient Greece Contributions to the Scientific Revolution

Building on the discoveries and knowledge of previous civilizations in Egypt, Mesopotamia, and Persia, among others, the Ancient Greeks developed a sophisticated culture immersing themselves with philosophical, mathematical, and scientific studies. One of the key points of Ancient Greek philosophy was the role of reason and inquiry. It emphasized logic and championed the applications of impartial and rational ideas for the natural world. Its general scientific and philosophical course however wasn't without certain faults and misdirection. Most notably was the lack of emphasis on the experimental and empirical approach to the physical world. One glaring example of their erroneous science was the dogma of celestial perfection and immutability, although this mistake may be blamed on the ancient Greek mythology and religion, believing in many gods and goddesses who were all supposed to reside up in the sky.

Another example of their misguided or imperfect approach to science, was the Greek's scientific quest to find what might be called 'Classical Elements', which were proposed to explain the nature and complexity of all matter in terms of simpler substances—a sort of ancient or primitive 'chemistry' or rather 'pre-alchemy'. Empedocles (around 494– 434 BC) was a Greek pre-Socratic philosopher. He originated the theory of the four classical elements, without performing any chemical experiments with substances, relying only on 'pure meditation' and perhaps some sort of intuition as well. Empedocles proposed the elements Fire, Air, Water, and Earth, and postulated that the differences of the structure of material in the world are produced according to the different proportions in which these four indestructible and unchangeable elements are combined with each other. Empedocles, like the atomists, stated that nothing new comes or can come into being; and that the only change that can occur in substances is a change in the specific combination of elements with elements.

The four elements of the Greeks were mostly qualitative aspects of matter, not quantitative, as in our modern elements in chemistry now. This theory of the four elements became the standard dogma in and around Europe for the next two thousand years, and later alchemists extensively developed this concept. Plato, according to his student Aristotle, added a fifth element, namely the quintessential, heavenly,

A. E. Kossovsky, *The Birth of Science*, Springer Praxis Books, https://doi.org/10.1007/978-3-030-51744-1_4

and incorruptible Aether, while classifying the other four elements as earthly and corruptible. Modern chemistry was certainly inspired directly by these Greek philosophical concepts, which at the end proved to be scientifically correct, although in a very different manner, with the atomic elements of the Periodic Table serving as its basis, and then with the more profound basis involving proton-neutron-electron combinations.

The Greeks made significant and original contributions to mathematics and geometry, but much less so to science, except in astronomy, in which they greatly excelled. Ancient Greek mathematicians such as Pythagoras, Euclid, and Archimedes—regarded nowadays with awe and admiration—are considered as the true 'initiators' or 'pioneers' of mathematics and especially geometry. Geometry is Greek!

Euclid of Alexandria (born around 325 BC) is referred to as the 'father of geometry'. The city of Alexandria was founded by Alexander the Great in 331 BC in passing through Egypt, just a few years before Euclid's birth. Euclid's book 'The Elements' is one of the most influential works in the history of mathematics, serving as the main textbook for teaching mathematics (especially geometry, but also conic sections and number theory) from the time of its first publication in antiquity until the late 19th or early 20th century.

Archimedes of Syracuse (around 287–212 BC) was a Greek mathematician, physicist, engineer, inventor, and astronomer. Archimedes is regarded as one of the leading scientists in classical antiquity. Archimedes is also considered the greatest mathematician of antiquity and one of the greatest mathematicians of all time. Archimedes anticipated modern calculus and analysis by applying concepts of infinitesimals and the method of exhaustion to derive and rigorously prove a range of significant geometrical theorems, including the area of a circle, the surface area and volume of a sphere, and the area under a parabola. He derived an accurate approximation of pi π, and created a system using exponentiation (powers) for expressing very large numbers. Archimedes was also one of the first to apply mathematics to physical phenomena. It is noted that Galileo praised and admired Archimedes, and referred to him as a 'superhuman'. Interestingly, some of Galileo's work conceptually resembles some of Archimedes's approaches. Archimedes died during the Second Punic War, when Roman forces captured the city of Syracuse after a two-year-long siege. He was killed by a Roman soldier despite explicit orders from General Marcus Claudius Marcellus that he should not be harmed.

Hippocrates of Kos (around 460–370 BC), was the most famous physician in antiquity. Hippocrates is considered one of the most outstanding figures in the history of medicine, and he is often referred to as the 'Father of Medicine' in recognition of his lasting contributions to the field; the many medical treatises that he wrote; as well as for his systematic and empirical investigation of diseases and remedies. The Hippocratic Oath, a medical standard for doctors, is named after him.

Claudius Ptolemy (around 100–170 AD) was a Greek mathematician, astronomer, and geographer, who lived in the city of Alexandria in the Roman province of Egypt, under the rule of the Roman Empire long after the fall of Greece. Ptolemy is famously known as the author of the book 'Almagest' which is a mathematical and astronomical treatise on the apparent motions of the stars and planetary paths. The Almagest is considered as one of the most influential scientific texts of all time, as it canonized a geocentric model of the universe. The Almagest cataloged the positions and timings of the stars and the planets under the framework and the assumption that everything revolves around the Earth. Almagest is also a key source of information about ancient Greek astronomy, serving as a sort of an encyclopedia of the sky for generations.

Anaxagoras of Clazomenae (born about 499 BC), was an astronomer who determined the cause of eclipses as positional alignments.

Democritus of Abdera (around 460–370 BC), realized that the Milky Way was composed of many millions of stars.

For Kepler, Galileo, Descartes, and Newton, all the many abstract discoveries of the ancient Greeks in geometry and mathematics were indispensable, directly influencing their progress, and indeed forming part of the basis of their work. The efforts, quests, and trials of the ancient Greeks in the physical sciences on the other hand bore no immediate fruits or substantial results to aid directly the scientific work of the 16th and 17th centuries, except in astronomy, where the careful and methodical observations and recordings of the sky by the ancient Greeks greatly influenced and partially paved the way for the work of Copernicus, Brahe, and Kepler.

Let us now turn our attention more specifically and at length to each of the six principle personalities of the 16th and 17th centuries, who collectively and sequentially led to the birth of science with the discovery of the true laws of motion and universal gravitation.

Chapter 5
Nicolaus Copernicus and the Heliocentric Model

Nicolaus Copernicus (1473–1543) was a Renaissance era mathematician and astronomer from Prussia (then part of Poland) who formulated an astronomical model with the Sun rather than the Earth at the center, meaning that the planets as well as Earth revolve around the Sun. This is called 'The Heliocentric Model'. Copernicus' work most likely was inspired by the Greek astronomer and mathematician Aristarchus of Samos who lived around 270 BC (nearly two millennia before him!) and who seems to be the first ever to propose such an astronomical model. Aristarchus himself was probably influenced by the Greek Pythagorean and pre-Socratic philosopher Philolaus of Croton who lived around 430 BC. Figure 5.1 depicts a portrait of Nicolaus Copernicus.

The publication of Copernicus' model in his book 'De Revolutionibus Orbium Coelestium' ('On the Revolutions of the Celestial Spheres'), just before his death in 1543, was a major event in the history of science, triggering the Copernican Revolution and making an important contribution to the Scientific Revolution. The neat and simpler Copernican heliocentric model was a tremendous improvement over the messy geocentric model of Ptolemy with its very complex system of epicycles and deferents, yet, it was not entirely the correct model. Ancient astronomers were overwhelmed by Aristotle's notion of circular motion for all the heavenly bodies, since the circle can be considered as the most perfect geometrical form in terms of its simplicity or in esthetic sense, as opposed to the more complex and 'messier' curves such as ellipses, ovals, parabolas, and such, and Copernicus could not let go of the assumption of circular motion of the planets.

According to Einstein, the bitter quarrel between Ptolemy and Copernicus has never really been settled yet, and in fact it could never be resolved in the future—on principle. Special and General Relativity hold that there exists no absolute motion nor even absolute acceleration, and questions such as who is moving and who is at rest are all relative to the subjective observer. The only thing that matters is that all the objective scientists—based on whatever moving or accelerating frames of reference—observe the same invariant laws of nature.

A. E. Kossovsky, *The Birth of Science*, Springer Praxis Books, https://doi.org/10.1007/978-3-030-51744-1_5

Fig. 5.1 Nicolaus
Copernicus

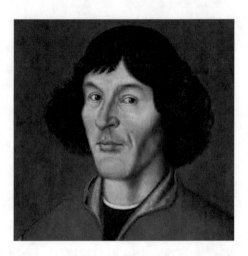

One important advantage of applying the Copernican heliocentric model is that it is much easier to work with when wider perspective is applied, with the stars of the Milky Way Galaxy serving as the inertial frame of reference for all the parts of the Solar System. This is so since the Sun can be considered as nearly stationary within the galaxy for 'short' time intervals such as a century or millennium. Ptolemy's geocentric model on the other hand is based on the non-inertial accelerating frame of reference of the rotating and revolving Earth as compared to the whole of the galaxy, and which is by far more difficult and messier to work with. Ultimately, even the galaxy itself rotates relative to the Local Group of Galaxies, and our group itself flies away very fast outward from the rest of the galaxies in the Universe.

Chapter 6
Tycho Brahe—The Meticulous Celestial Recorder

Tycho Brahe (1546–1601) was a Danish/Swedish aristocrat—heir to several of Denmark's most influential noble families. He was an astronomer as well as astrologer and alchemist. He is known for his accurate and comprehensive astronomical observations—creating and applying more accurate instruments of measurement. He was the last of the major naked-eye astronomers, working without the telescope for his observations. He has been described as "the first competent mind in modern astronomy with the passion for exact empirical facts". His astronomical observations were some five times more accurate than the best available observations at the time. Figure 6.1 depicts a portrait of Tycho Brahe.

Tycho's initial fame in astronomy stemmed from his observations and subsequent writing about the supernova of 1572, which at its peak was as bright as Venus, and it was visible in the daytime for about two weeks. This observation strongly contradicted the ancient Aristotelian belief in celestial immutability. Supernova is the spectacular and explosive death of a super massive star weighing approximately 8–15 solar masses. Supernova explosions happen rarely. In our own galaxy, the Milky Way, the last supernova happened in the year 1604, and it is known as Kepler's Supernova, since Kepler described it in his 1606 book titled 'De Stella Nova in Pede Serpentarii' ('On the New Star in the Foot of the Serpent Handler').

Tycho's publication of the book 'De nova stellar' ('The New Star') led King Frederick II of Denmark to offer Tycho the use of the island of Hven (located today between Sweden and Denmark) for the purpose of conducting astronomical observations, and provided him with considerable financial and material assistance. Tycho built an observatory he called Uraniborg in 1576, and with his large group of assistants made highly precise naked-eye observations. Tycho collected an astounding amount of astronomical data, writing down the location of stars and planets in the sky and the associated time of observations.

Uraniborg castle was the first ever building in the world designed primarily to be used as a place of astronomical observation and study, and was ornamentally designed with pictures of astronomers and inscribed poems about astronomy. Tycho

A. E. Kossovsky, *The Birth of Science*, Springer Praxis Books, https://doi.org/10.1007/978-3-030-51744-1_6

Fig. 6.1 Tycho Brahe

held many parties and meetings at Uraniborg castle, inviting scientists and others, and often with heavy drinking involved.

Unfortunately for Tycho, after the death of King Frederick II and the coronation of the young prince-elect in 1596, not only did he totally lose the support of the royal court, but he also was embroiled in conflicts and then even accused of heresy. Finally in 1597 he decided to leave Denmark, staying with friends in his self-imposed exile in Germany.

Luckily for Tycho he was soon warmly invited to Prague in 1599 by Rudolf II, Holy Roman Emperor; was treated very well and kindly; and was assisted in building his second observatory in a castle near Prague. He worked under Emperor Rudolf II sponsorship for the last three years of his life until his death at the end of 1601.

As soon as Tycho settled in Prague, he invited Johannes Kepler who came there on February 4, 1600, and met with Tycho for the first time. This was a very fortunate occurrence for science, or rather a very lucky coincidence, for the young 29-year old Kepler was not yet a major figure in either mathematics or astronomy, except for publishing his first astronomical work, 'Mysterium Cosmographicum' ('The Cosmographic Mystery') in 1596. But Tycho needed an assistant, someone with mathematical talents and a good knowledge in astronomy, and he expected that Kepler would help him with the mathematical elaboration of his data and especially in using the data to prove Tycho's own model in astronomy. Tycho's proposed model of the Solar System was an awkward model which now strikes us as an ugly hybrid, and it supposes that the Sun and Moon orbit the Earth, but that the other planets orbit the Sun. Tycho's model can be thought of as the last attempt to place the Earth in a favored position (as the only unmoved object) while allowing the Sun to be the center of some motions such as the planets. Tycho's astronomical model had suited Denmark's Protestant church in a religious sense, since Martin Luther—the founder of the Protestant doctrine—had rejected the views of Copernicus (who lived at about the same time as Luther) on Biblical and religious grounds. About a century later, the

Catholic Church (Luther's nemesis) followed in Luther's footsteps and stringently rejected Copernicus astronomical views, culminating in the Trial of Galileo in 1633.

Over the next two months Kepler stayed as Tycho's guest, analyzing some of Tycho's observations of Mars. Although Tycho was very guarded and secretive with his data, Kepler impressed him such that soon he was given a bit more access to Tycho's Mars-related data. Tycho's observations regarding measurements of Mars did not agree with either Ptolemy's geocentric model or Copernicus' heliocentric model, and this discrepancy between theory and data foreshadowed a major revolution in astronomy. Kepler was particularly focused on Tycho's data about Mars, which he thought he could use to test some of his own theories about planetary motion, but he estimated that the work would take at least two years—since he was not allowed to simply copy Tycho's data for his own use, and this hindrance considerably retarded his progress.

The relationship between Tycho and Kepler was often tense and conflicted, and they had several angry exchanges and disagreements. After a series of ups and downs, they came to an agreement about permanent position, and Kepler came to settle in Prague in 1600 and began working with some of Tycho's data.

Tycho's selection of Kepler as his chief mathematical assistant was a momentous and very significant decision that greatly impacted the course of history. Any other selection for the work would almost certainly not yield for science the three planetary laws discovered by Kepler. Somehow, as if by magic, the most capable person around Europe, or rather perhaps the *only* person capable of applying Tycho's accurate data by distilling and consolidating it into general ideas and formulas, was properly selected for the job! Tycho without Kepler would yield only raw data without an analysis, and Kepler without Tycho would grant us an exceptional and innovative mind, but one that would be unexpressed and wasted—with nothing to work on.

Their collaboration lasted only about two years. Tycho suddenly contracted some bladder or kidney ailment after attending a long and formal party in Prague where he was reluctant to relieve himself in the restroom as a courtesy to the hosts, and he died eleven days later, on 24 October 1601, at the age of 54. The night before Tycho died he suffered from a delirium during which he was frequently heard to exclaim that he hoped he would not seem to have lived in vain, then urging Kepler to finish his work and expressing the hope that Kepler would do so by adopting Tycho's own planetary system, rather than that of Copernicus. A team of Czech and Danish scientists collecting Tycho's bone, hair and clothing samples for analysis reported their conclusion in November 2012 that Tycho most likely died of a burst bladder.

Before Tycho died he gave all his data to Kepler and suggested that Kepler would become his successor. Soon afterwards, Kepler was appointed his successor as imperial mathematician with the responsibility to complete Tycho's unfinished work. The next 11 years as imperial mathematician would prove the most productive of Kepler's life—particularly due to having total and free access to Tycho's astronomical data, and working much more freely and independently.

Here fate intervened again. Had Tycho lived to the ripe old age of, say 70 or 80, Kepler might not have been able to discover the three planetary laws. Tycho needed to call Kepler in, only Kepler, nobody else, and then to die soon afterwards, leaving his accurate data for Kepler alone, to analyze and brood over freely. That's how science triumphed here in this particular story.

Tycho led a colorful life. At age 20 Tycho lost part of his nose in a sword duel against a fellow Danish nobleman, and which resulted in Tycho losing the bridge of his nose and having to wear a prosthetic nose made of gold and silver for the rest of his life.

Tycho fell in love with a daughter of a Lutheran minister, who was a commoner. Tycho never formally married her, since if he did he would lose his noble privileges. They lived together as unmarried and informal couple. Each would however maintain their social status, and any children they had together would be considered commoners. Many of Tycho's noble family members strongly disagreed with this romantic relationship involving a commoner, and many churchmen would continue to hold the lack of a divinely sanctioned marriage against him. Together they had eight children, six of whom lived to adulthood.

Chapter 7
Johannes Kepler—The First Data Analyst

Johannes Kepler (1571–1630) was a German mathematician, astronomer, astrologer, and a key figure in the seventeenth century Scientific Revolution. Born prematurely to a relatively poor German protestant family, he was said to be weak and sickly as a child. His love for astronomy grew early in life when he observed the great comet of 1577 and the lunar eclipse in 1580. However, a childhood smallpox infection left him with weak vision and crippled hands, limiting his ability in the observational aspects of astronomy. After moving through grammar school, Latin school, and seminary, Kepler attended the University of Tubingen, where he studied philosophy and theology. Despite his desire to continue studying theology and to become a minister, near the end of his studies, Kepler was recommended for a position as teacher of mathematics and astronomy at the Protestant school in Graz. He accepted the position in April 1594, at the age of 23. Figure 7.1 depicts a portrait of Johannes Kepler.

Teaching of the Copernican heliocentric model was at the time prohibited by the Church, but his teacher, who was a strong believer of the heliocentric model, used to teach it secretly to his brilliant students, such as young Kepler, and thus Kepler learned both, the Ptolemaic and Copernican systems of planetary motion. Sensing intuitively that the Copernican model of the universe is the correct one, Kepler made it his mission in life to prove rigorously and beyond any doubt that the heliocentric model is the correct one. Kepler together with his contemporary Galileo, made multiple assaults on the Ptolemaic system, and successfully finalized the astronomical revolution in favor of the Copernican model. Kepler spent over 25 years analyzing Tycho's data and attempting to unlock the secrets it contained. In attempting to find order in this data Kepler was led down many failed paths. His overly creative mind at times caused him to waste time and efforts on wrong ideas. His first astronomical conjecture was that the orbits of the six planets were somehow related to the five perfect geometric solids known to the Greeks. They were not! He spent years analyzing patterns in music looking for the 'harmony of the spheres' as if proportions of the Solar System were connected to musical scales. They were

A. E. Kossovsky, *The Birth of Science*, Springer Praxis Books, https://doi.org/10.1007/978-3-030-51744-1_7

Fig. 7.1 Johannes Kepler

not! Kepler was not afraid to 'look outside the box' and test new and unconventional thoughts. He eventually found success formulating three seminal laws regarding planetary motion.

This author would like to suggest that Kepler should be considered the first ever data analyst, albeit without the more modern use of Descriptive Statistics, Mathematical Statistics, and their numerous fancy tools, such as outliers analysis, correlation and regression, measures of skewness, and so forth. In all the time of antiquity before Kepler, there was not a single instance when someone was analyzing an existing physical set of data for the purpose of synthesizing it, or for the purpose of finding some patterns; and certainly not for the purpose of fitting the data into some exact equations or formulas relating variables.

The story of Johannes Kepler is remarkable because in him we have a mystical figure who was driven by aesthetic and spiritual considerations, but who nonetheless placed agreement with observations above all such considerations. This uncompromising position led, after many twisting years, to one of the most remarkable breakings of a mental block in the history of science, namely the overcoming of the ancient Greeks' misleading belief in perfect circular planetary paths, and to the correct selection of the ellipse as the geometric shape for their orbits.

Chapter 8
Kepler's Attempts to Fit Orbits Geometrically

In 1596 at the age of 26 Kepler published his first major astronomical work, **'Mysterium Cosmographicum'** ('The Cosmographic Mystery'). This was the first ever published defense of the Copernican system. It should be noted that part of Kepler's enthusiasm for the Copernican system in general stemmed from his theological convictions about the connection between the physical and the spiritual. His first manuscript of Mysterium contained an extensive chapter reconciling heliocentrism with biblical passages that seemed to support geocentrism. Kepler was very much concerned with religion in general, and he did not wish to contradict his religious beliefs with scientific ideas.

The inspiration and prelude to Mysterium was Kepler's sudden [but false] insight, conjecturing that regular polygons which bound inscribed circles and circumscribed circles could point to definite ratios of the radii of these surrounding circles, and might be the geometrical basis of the Solar System. While teaching an astronomy class in Graz on July 19, 1595, Kepler had drawn for his students the circumscribed and inscribed circles for an equilateral triangle as depicted in Fig. 8.1. Instantly Kepler had the insight that this could be very significant to astronomy, realizing that the ratio of the radii for the circumscribed and inscribed circles was nearly the same as the ratio of the orbital radii of Jupiter and Saturn; and he initially could not believe that this equivalency was merely a coincidence.

Regular polygons are perfectly symmetrical two-dimensional shapes using N straight lines—all of the same length, and all connected via an identical-valued angle. For N = 3 it's the equilateral triangle, for N = 4 it's the square, for N = 5 it's the symmetrical pentagon, and so forth.

Figure 8.1 depicts the construction for the two circles inscribed and circumscribed around an equilateral triangle. Basic Geometric and Trigonometric calculations of this construct show that the ratio of the inner inscribed circle to the outer circumscribed circle is always 1/2, and that this is so regardless of the size of the chosen equilateral triangle. Kepler was intrigued by the fact that this ratio corresponds approximately to the ratio of Jupiter's radius (779 million kilometers) to Saturn's radius (1434 million kilometers) which is 0.543 and quite close to 1/2! In addition,

A. E. Kossovsky, *The Birth of Science*, Springer Praxis Books, https://doi.org/10.1007/978-3-030-51744-1_8

Fig. 8.1 Circles inscribed
and circumscribed around a
triangle

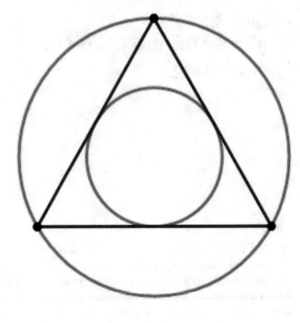

Fig. 8.2 Circles inscribed
and circumscribed around a
square

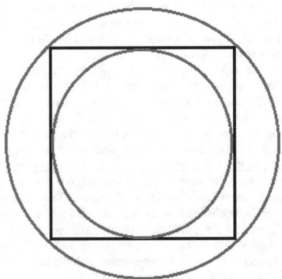

this 1/2 ratio is quite close to the ratio of Mercury's radius (58 million kilometers)
to Venus' radius (108 million kilometers) which is 0.537!

Figure 8.2 depicts the arrangement for the square—being the second simplest
regular polygon—and there the ratio of the inner inscribed circle to the outer circumscribed circle is $1/(\sqrt{2})$ or $1/(1.414)$, namely 0.707, so that here the circles are more
similar to each other, and differences in sizes are less dramatic as compared with

the arrangement of Fig. 8.1 of circles around a triangle. Applying this ratio to the ratio of Venus' and Earth's radii also gives an excellent fit and suggests that perhaps the model is workable; since the ratio of Venus' radius (108 million kilometers) to Earth's radius (150 million kilometers) turned out to be 0.720 approximately, which is quite close to 0.707!

The general formula for ratios of the radii of inscribed and circumscribed circles to an N-sided regular polygon is straightforward to derive, and the result is simply stated here without detailed derivation. If the circumscribed circle has radius C, and the inscribed circle has radius I, then the ratio of the radii is **I/C = cos(180°/N)**. Applying this result for $N = 3$ to $N = 13$ we obtain the following table:

N	I/C
3	0.500
4	0.707
5	0.809
6	0.866
7	0.901
8	0.924
9	0.940
10	0.951
11	0.959
12	0.966
13	0.971

Kepler hoped to somehow match values within this set of ratios of abstract circles with the set of ratios of the real (supposed) circles of the planets. Surely the whole arrangement and the conjecture is based firmly on the Copernican model where the planets revolve around the Sun. Kepler's enthusiasm was driven by the thought that a confirmation of this conjecture would surely support if not totally validate the heliocentric model. No one before Kepler ever inquired about a rationale for the way the planets are arranged sequentially in their particular orbits, let alone attempt to actually answer this by providing an exact numerical and geometrical explanation!

One simplistic attempt is depicted in Fig. 8.3 where three polygons are arranged in their natural order regarding the number of sides, and where the equilateral triangle is inscribed by one circle as well as circumscribed by another circle, all of which in turn is circumscribed by a square, which is then also circumscribed by a circle, all of which in turn is circumscribed by a pentagon, which is then also circumscribed by a circle. Continuing this scheme to hexagon and heptagon (not shown in Fig. 8.3) would lead to six such geometrically-derived ideal circles representing all existing six planets known at the time.

Fig. 8.3 Circles inscribed and circumscribed around polygons

In case this particular natural arrangement did not work out, Kepler was ready to test a model in which the polygons are in an inverted order, beginning with heptagon at the inner-most part, and ending with a triangle at the outer part. He was even ready to accept a model without any particular order for the regular polygons. More-over, Kepler was willing to accept a repeated use of any polygon, and in any order whatsoever, if such arrangement could somehow fit the actual astronomical data on hand.

But Kepler's initial enthusiasm about this conjecture soon subsided after failing to find a unique arrangement of regular polygons that could fit known astronomical observations (even with extra imaginary planets added to the system). For the partic-ular arrangement of Fig. 8.3, the circles tend to be approximately equally spaced from each other, and therefore ratios are increasing from 0.50 to 0.901; in contrast, actual distances between the great circles of the planets' orbits are increasing and their ratios are approximately 'stuck' in the range of 0.3 to 0.7. Summarizing the five ratios of the circles of the six planets we get:

The ratio of Mercury's radius (58 million kilometers) to Venus' radius (108 million kilometers) is **0.537**.

The ratio of Venus' radius (108 million kilometers) to Earth's radius (150 million kilometers) is **0.720**.

| **Tetrahedron** | **Cube** | **Octahedron** | **Dodecahedron** | **Icosahedron** |
| **4 Faces** | **6 Faces** | **8 Faces** | **12 Faces** | **20 Faces** |

Fig. 8.4 The five polyhedrons known as platonic solids

The ratio of Earth's radius (150 million kilometers) to Mars radius (228 million kilometers) is **0.658**.

The ratio of Mars' radius (228 million kilometers) to Jupiter's radius (779 million kilometers) is **0.293**.

The ratio of Jupiter's radius (779 million kilometers) to Saturn's radius (1434 million kilometers) is **0.543**.

Even a cursory comparison between this set of five planetary ratios to the set of eleven ratios of inscribed and circumscribed circles around regular polygons given earlier—leads to the conclusion that they do not really correspond. Yet, in spite of his profound disappointment, Kepler persisted, and then he suddenly had another insight. The six planets do indeed orbit the sun as in around a flat plane, however the world is three dimensional, hence instead of applying polygons which are planar figures in two dimensions, perhaps three-dimensional analogs of regular polygons could be applied with better results. Kepler then gave up on polygons and began experimenting with 3-dimensional symmetrical solids instead.

There were only **SIX** known planets at that time, Mercury, Venus, Earth, Mars, Jupiter, and Saturn.

The **FIVE** Platonic symmetrical solids (convex polyhedron) have been known since antiquity. They are constructed by congruent (identical in shape and size) regular (all angles equal and all sides equal) polygonal faces with the same number of faces meeting at each vertex. In other words, polyhedrons are solids in three dimensions with flat polygonal faces, straight edges and sharp corners or vertices. They are totally symmetric in any sense of the word. There exist **only five** such perfectly symmetrical solids. They are named 'Platonic solid' for the ancient Greek philosopher Plato who misguidedly associated them with the equally misguided five classical elements—earth, water, air, fire, and aether—believed in antiquity to constitute or to explain the nature and complexity of all matter in terms of these five simpler substances. Figure 8.4 depicts those five Platonic solids, named tetrahedron, cube, octahedron, dodecahedron, and icosahedron.

For Kepler, this numerical correspondence became quite interesting; for *six planets*, there will be five intervening spaces, to be occupied by five platonic solids—all circumscribed or inscribed by *six spheres* representing the orbits of the planets! "What a perfect match! Surely this is not just a meaningless coincidence"—Kepler thought. Figure 8.5 depicts Kepler's own drawing of the arrangements of spheres inscribed and circumscribed around Platonic solids.

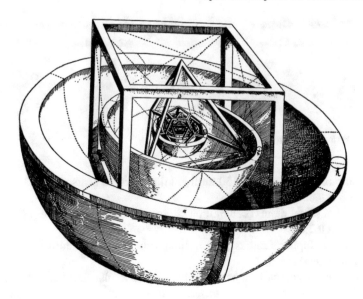

Fig. 8.5 Platonic solids inscribed and circumscribed by spheres

Euclid devoted the last book XIII of the Elements to these regular solids, which serves in a sense as the crowning achievement to his geometry, presenting a proof that only five solids meet those strict symmetry criteria: Tetrahedron, Cube, Octahedron, Dodecahedron, and Icosahedron. This was the first known proof that exactly five regular polyhedra exist.

By ordering the solids selectively—octahedron, icosahedron, dodecahedron, tetrahedron, cube—as opposed to their natural order according to the number of faces, Kepler found that the spheres could be placed at intervals corresponding to the relative sizes of the planets' orbits—assuming the planets circle the Sun.

The table in Fig. 8.6 depicts the radii of the inscribed and circumscribed spheres around polyhedrons, and it is a summary of the long and very tedious calculations and derivations in Solid Geometry. In addition, the table includes a column showing the ratios of (inscribed)/(circumscribed) radii for these spheres. The measures of the radii of the inscribed and circumscribed spheres are given in terms of the dimension S which measures the side length of a regular polygon or edge length of a Platonic solid.

Figure 8.7 depicts the approximate correspondence between the ratios of the planets' orbits and the ratios of the inscribed and circumscribed spheres around this particular ordered set of polyhedrons. It is probable that Kepler has scripted and stared at such a table himself with great joy and satisfaction. Figure 8.8 depicts the bar chart of the two sets of ratios which appear to correspond in the approximate.

To the perfectionist, there are only two minor issues here to ponder. The first issue is the consistent tiny advantage of the ratios of the spheres of the Platonic solids compared with the ratios of the orbits of the planets, as the former set is always a bit

SOLID	Inscribed Radius	Circumscribed Radius	RATIO
octahedron	S((√6)/6)	S((√2)/2)	0.577
icosahedron	S((3√3 + √15)/12)	S((√(10 + 2√5))/4)	0.795
dodecahedron	S((√(250 + 110√5))/20)	S((√15 + √3)/4)	0.795
tetrahedron	S((√6)/12)	S((√6)/4)	0.333
cube	S(1/2)	S((√3)/2)	0.577

Fig. 8.6 Sizes and ratios of inscribed and circumscribed spheres

Fig. 8.7 Correspondence between ratios of planets and spheres

Planet	Ratio	Ratio	Polyhedron
Mercury			
	0.537	0.577	octahedron
Venus			
	0.720	0.795	icosahedron
Earth			
	0.658	0.795	dodecahedron
Mars			
	0.293	0.333	tetrahedron
Jupiter			
	0.543	0.577	cube
Saturn			

Fig. 8.8 Bar chart of the correspondence between the ratios

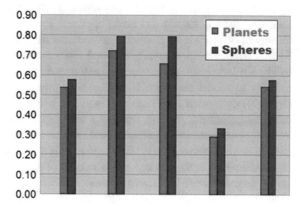

higher than the latter set. The second issue is the need to scramble and disturb the natural order of these Platonic solids in terms of the number of faces. If nature behaves so elegantly, symmetrically and geometrically, why did it not attempt to achieve that extra measure of perfection by applying just a little more effort—organizing the solids in their natural order?! Nonetheless, this was surely a great feat by Kepler! Here we have exactly **6 spheres** around these unique set of five solids to serve as a model and to be compared with exactly **6 orbits** of the known planets at the time; *moreover*, this comparison yields a very close agreement between the two sets of 5 ratios!

This apparent correspondence between the orbits of the planets and the spheres around Platonic Solids is the main theme of Kepler's book 'Mysterium Cosmographicum', where he proclaims: "You now have the reason for the number of the planets."

His model, however ingenious, did not agree well with observations, particularly in light of subsequent, more accurate, observational data. Indeed, his model was entirely disapproved by posterior discoveries of the planets Uranus and Neptune, since there are no additional Platonic solids that could determine their distances from the Sun! Now we know that there are no compelling geometrical reasons for the planets to be at the distances they are from the Sun. Modern astronomy envisions planets and stars forming randomly and chaotically under the forces of gravity and inertia from the gas and dust in intergalactic space, with no relation to such particular geometry whatsoever.

However, the book made his reputation as an imaginative thinker and keen astronomer. More importantly, this was one of the first hypotheses based on a heliocentric model. It brought Kepler to the attention of the scientific community, including Galileo Galilei and Tycho Brahe.

The rest of Kepler's entire astronomical career was simply an elaboration and extension of the questions he sought to address in this book: Why were there exactly six known planets in the Solar System? Why are they spaced around the Sun in that particular manner? How do they move exactly in relation to time, and what are their speeds? His radical and novel approach in answering these questions was to create out of thin air an elaborate geometrical model for the structure of the Solar System based on spheres inscribed and circumscribed around five nested Platonic solids. Kepler's imaginative approach in borrowing from geometry shapes and results in order to model the Solar System was highly creative! No one has ever (even remotely) performed syntheses of this nature before, either in astronomy, or in any other scientific field! In many ways, this book represented a turning point for the way astronomy and science in general would be done in the future, for it sought to provide reasons behind what had heretofore been mainly a descriptive science. Kepler scholar Owen Gingerich has remarked in his account of Kepler's life and work: "Seldom in history has *so wrong* a book been *so seminal* in directing the future course of science". Modern astronomy owes much to Mysterium Cosmographicum, despite the flaw in its main thesis. The book also represents the first major step in establishing and organizing the heliocentric model via a mathematical model, at the expense of the geocentric model.

Though the details of his geometric model would be modified in light of his later work, Kepler never relinquished the Platonist polyhedral-spherist cosmology of Mysterium Cosmographicum. His subsequent main astronomical works were in some sense only further developments of it, concerned with finding more precise inner and outer dimensions for the spheres, and the relations between the sizes of the orbits to time and speed.

Significantly, Mysterium Cosmographicus contains also a philosophical-mechanical point of view of the Solar System that would consequently be proven true by Newton, and most likely was a key inspiration for Newton's formulation of his law of universal gravitation. Kepler is simply stating this vista without analyzing any astronomical data whatsoever, and without any mathematical calculations. Here Kepler took the first tentative step toward the modern picture, where the Sun—by its gravitation—controls the motions of the planets. In the first edition, Kepler attributed to the Sun a motricem animam ("moving soul"), which causes the motion of planets. In Aristotelian cosmology, each of the heavenly bodies had its own individual soul, which produced and guided their motion. Kepler explicitly rejected the idea, and suggested instead that in the Solar System there is but a single soul, and it is in the Sun. In the second edition of Mysterium Cosmographicum Kepler came closer to the modern picture, as he supposed that some force—which, like light, is "corporeal" but "immaterial" emanates from the Sun and drives the planets—and discarding the idea of 'souls' altogether. Significantly, Kepler sought a reason why the planets further from the Sun were slower, and he speculated that the Sun is the source of the planets' motion, just as it was the source of light. He attributed to the Sun a certain force whose influence weakens at greater distances, just as strength of sunlight grows lesser at greater distances. While not explicitly stating that the force is *attraction*, with arrows or vectors of forces pointing from planets to the Sun, Kepler is the first thinker to hint at universal gravitation. In this sense, the Copernican heliocentric revolution played a very direct and key role in the formation of modern physics, far beyond the immediate and more 'provincial' need to understand our territorial orientation within the Solar System and its structure.

Chapter 9
Kepler's 1st Law—Discovery of Elliptical Orbits

In 1600 Tycho Brahe invited Kepler to Prague. After Tycho's death in 1601 Kepler was appointed his successor as imperial mathematician at the royal court. Now, being in the possession of Tycho's entire astronomical data and with a strong focus on Mars, Kepler would make his most significant discoveries which would prove to be true and lasting.

Mars was to Kepler what Mercury would be to Einstein, but in reverse theoretical/observational order. For Einstein's General Relativity, the *deductive* method was used to create a whole new theory from abstract postulates (without consideration of data) and which would be dramatically and triumphantly confirmed by Einstein himself soon afterwards via already available empirical data of observation about the perihelion precession of Mercury's orbit. For Kepler the *inductive* method was used via Tycho's empirical data on Mars' orbit, to synthesize and analyze the data, to get clues from the data, and then to arrive at his theoretical first and second laws of planetary motions.

Kepler began with an analysis of Mars, and calculated various approximations of Mars' orbit, but he was not satisfied with the outcome due to slight differences between the theoretical circular orbit and the actual data. Even though the size of the discrepancy was not that great, Kepler strived to obtain better fit between theory and data. Finally after many trials and errors, Kepler grasped the main problem of the heliocentric model—namely that the orbits of the planets are elliptical, not circular, and with the Sun at one focal point of the ellipse, away from its center. This revolutionary idea is called Kepler's first law. Till then, astronomers were stuck with Plato's idea that the heavenly bodies could only orbit in a circular path. It may not be quite obvious or well-known nowadays, but Kepler's first law of planetary motion was a giant leap in humanity's intellectual progress, boldly breaking away from two millennia of Greek dogma.

A. E. Kossovsky, *The Birth of Science*, Springer Praxis Books, https://doi.org/10.1007/978-3-030-51744-1_9

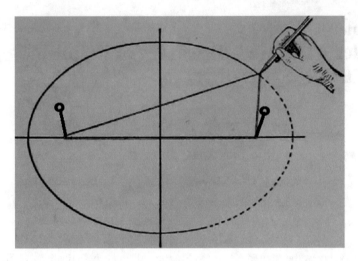

Fig. 9.1   The drawing of an ellipse with a string

Kepler's first law:

The orbit of a planet is an ellipse with the Sun at one of the two focal points.

An ellipse is an oval-shape or egg-shape closed curve. It can be thought of in a sense as an original circle squeezed simultaneously from above and from below.

Figure 9.1 illustrates the drawing of an ellipse with the use of two pins or two poles, a string, and a pencil, keeping the tension in the string constant all the way around.

If we let 1 unit on the Cartesian Plane equals 45,487,666 km, then the elliptical orbit of Mars with the Sun at one focal point is then depicted in Fig. 9.2.

Not shown in Fig. 9.2 is the other focal point on the right, and this is so because it does not signify anything physical relating to Mars orbit. In other words, the focal point on the right is missing because it is just a mathematical abstraction.

Even though Mars orbit is truly elliptical, its shape appears almost indistinguishable from a circle. Figure 9.2 was drawn correctly, extending 10.02 units on the major (horizontal) axis, and 9.98 units on the minor (vertical) axis, but such small discrepancy in the lengths cannot be visualized here easily, or perhaps not at all. In other words, if Mars's orbit was ever originally circular at the time of its creation, it must be then that Mother Nature who was quite affectionate with Mars for some reason had given it a very gentle, soft, and loving squeeze from above and from below when attempting to convert its orbit from circular into an elliptical one, thus altering its original circular shape only very slightly.

This delicate configuration might explain the challenges Kepler faced when trying to ascertain the true shape of the orbit of Mars. It could easily mislead astronomers to believe that the orbit is circular, because in reality it is almost exactly so. The

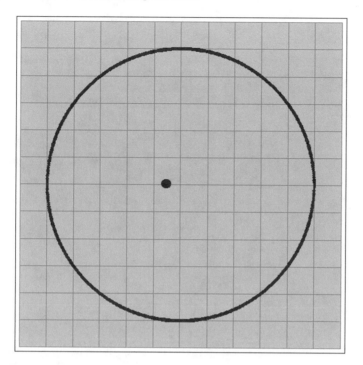

Fig. 9.2 Mars' elliptical orbit with the sun at a focal point

only clue that could possibly reveal its true elliptical nature is the position of the Sun which is moderately away from the center.

Kepler's first law is deemed to be the first ever case (followed by endless later cases) in the history of science when purely abstract mathematical concepts and results later find significant applications in the sciences, describing and explaining physical phenomena. Well, here the purely abstract geometrical study of Conic Sections of the ancient Greeks, mainly by Apollonius of Perga almost two millennia before Kepler, enabled Kepler to identify the precise shape of planetary orbits. Conic Sections are the curves obtained by the intersection of the surface of a cone with a plane, namely the hyperbola, the parabola, the ellipse, and the circle.

Chapter 10
The Eccentricity of Elliptical Orbits

The eccentricity of the ellipse is a measure of the deviation in the ellipse from that most symmetrical shape of the circle; namely by how much the perfect circle has been squeezed into that elliptical shape.

The numerical calculation in the definition of eccentricity is very intuitive and easily grasped. First we need to measure the widest line (horizontal perhaps) spanning from edge to edge, as well as the shortest line (vertical perhaps) spanning from edge to edge. These two lines form a cross inscribed inside the ellipse as shown in Fig. 10.2. There lines are called Major-Axis and Minor-Axis respectively.

The mathematical or numerical measure of eccentricity is simply the ratio (Minor-Axis)/(Major-Axis), namely the degree by which Minor-Axis is smaller than Major-Axis, although a few mathematical operations are applied to this ratio by squaring it, taking the difference between it and 1, and then taking the square root of the difference, but all these mathematical transformations are mere conveniences.

$$\text{Eccentricity} = e = \sqrt{(1 - (\text{Minor-Axis}/\text{Major-Axis})^2)}$$

If the ellipse is totally symmetrical, then it's not really an ellipse, but rather a circle, and therefore Minor-Axis = Major-Axis, implying that the ratio is 1, and therefore e is 0, so that there is no eccentricity whatsoever, everything is normal, namely symmetrical and circular.

If the ellipse was gently squeezed, so that say Minor-Axis is 90% of Major-Axis, the ratio is 0.9, and therefore e is 0.436.

If the ellipse was tightly squeezed, so that say Minor-Axis is only half of Major-Axis, the ratio is 0.5, and therefore e is 0.866.

A. E. Kossovsky, *The Birth of Science*, Springer Praxis Books, https://doi.org/10.1007/978-3-030-51744-1_10

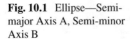

Fig. 10.1 Ellipse—Semi-major Axis A, Semi-minor Axis B

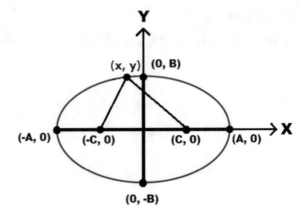

Fig. 10.2 Ellipse—Major-Axis is 2A and Minor-Axis is 2B

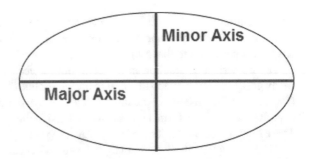

On the Cartesian Plane an ellipse with semi-major axis A and semi-minor axis B is the algebraic solution to the X and Y equation $\mathbf{X^2/A^2 + Y^2/B^2 = 1}$, assuming A > B.

An ellipse is also a curve in a plane surrounding two focal points such that the sum of the two distances to the two focal points is constant for every point on the curve. The two focal points on the Cartesian Plane are $(-C, 0)$ and $(+C, 0)$ where $\mathbf{C^2 = A^2 - B^2}$.

Figure 10.1 depicts the relevant points of the ellipse on the Cartesian Plane. Figure 10.2 depicts the definitions of 'major-axis' and 'minor-axis' as simply the maximum horizontal length and the maximum vertical length respectively, namely double of A and double of B. The expression 'Semi-Major Axis' means half of the Major Axis. It should be noted that the two focal points lie on the major (largest) axis.

On the Cartesian Plane, a circle with radius R is the algebraic solution to the X and Y equation $\mathbf{X^2 + Y^2 = R^2}$. The circle can be interpreted as a special case of the ellipse when A = B and then naturally each is given the same designation of the letter R. Therefore the elliptical equation for the circle can be written in a simpler form as $\mathbf{X^2/R^2 + Y^2/R^2 = 1}$, and then easily converted to the circular form of equation after the simultaneous multiplications of both sides of the equation by $\mathbf{R^2}$.

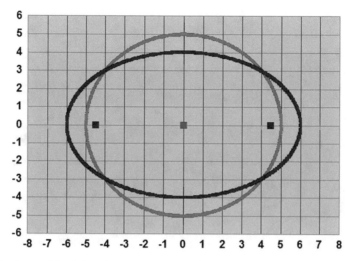

Fig. 10.3 Comparable circle and ellipse superimposed

Now that we have assigned two exact parameters A and B which precisely and fully describe the ellipse (C is not really a parameter since it is derived directly from A and B as seen in the previous page), eccentricity e can be succinctly defined in terms of A and B as $e = \sqrt{(1 - B^2/A^2)}$, or equivalently as $e = C/A$.

The value of eccentricity progresses from 0 for the perfect circle where A = B, to nearly the value of 1 in the limit for an ellipse arising from a highly squeezed circle—where A is much larger than B.

As it happened, even a tiny little gentle squeeze on a circle, where the newly created ellipse still looks very much like a circle almost, and where A is just slightly larger than B, causes the focal points to move considerably and noticeably away from the center. Alternatively stated, the two focal points of the circle are merged and fused into a singular point called the center, and they are **_highly sensitive_** to squeezing, separating from each other significantly even under the most gentle squeeze on the circle.

When a bit more squeezing occurs, and where the difference between A and B is just a bit larger, it causes the focal points to reside at quite extreme distances to the right and to the left, far away from the [original] center. An example of one such ellipse with A = 6 and B = 4 is depicted in Fig. 10.3, and superimposed by a circle of radius 5 for comparison. Eccentricity is calculated as $e = \sqrt{(1 - B^2/A^2)} = \sqrt{(1 - 4^2/6^2)} = \sqrt{(1 - 0.444)} = 0.745$, and the focal point distance from the center is $C = \sqrt{(A^2 - B^2)} = 4.47$, meaning that the focal points are quite near the edges.

As another example, Fig. 10.4 depicts an ellipse with only a minor difference between A and B, and where A = 5.5 and B = 4.5. Eccentricity is calculated as $e = \sqrt{(1 - B^2/A^2)} = \sqrt{(1 - 4.5^2/5.5^2)} = \sqrt{(1 - 0.669)} = 0.575$, and the focal point distance from the center is $C = \sqrt{(A^2 - B^2)} = 3.16$, meaning that the two focal

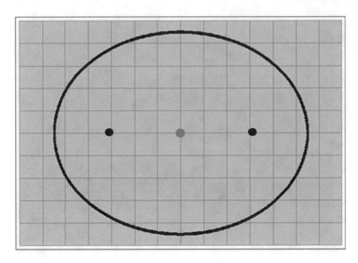

Fig. 10.4   An ellipse, A $= 5.5$, B $= 4.4$, focal point $= 3.16$

points are quite far away from the [original] center - in spite of the relatively minor 1 unit difference between the values of A and B, namely $5.4 - 4.5$.

Mars' eccentricity is of the very low value of 0.0934.

For Mars, B is only very slightly less than A.

A $=$ Semi-major axis is 227,936,637 km.

B $=$ Semi-minor axis is 226,939,986 km.

C $=$ Focal point $= \sqrt{(A^2 - B^2)} = 21{,}292{,}093$ km.

$$e = \sqrt{(1 - B^2/A^2)} = \sqrt{(1 - 0.99127)} = \sqrt{(0.00873)} = 0.0934.$$

If we let 1 unit on the Cartesian Plane equals 45,487,666 km, then the dimensions of Mars orbit converts into:

A $=$ Semi-major axis is 5.01 units.

B $=$ Semi-minor axis is 4.99 units.

C $=$ Focal point $= 0.47$ units.

$$e = \sqrt{(1 - B^2/A^2)} = \sqrt{(1 - (4.99^2)/(5.01^2))}$$
$$e = \sqrt{(1 - 0.99127)} = \sqrt{(0.00873)} = 0.0934$$

The elliptical orbit of Mars with the Sun at one focal point was depicted earlier in Fig. 9.2.

Figure 10.5 depicts the superimposed orbits of Earth and Mars.

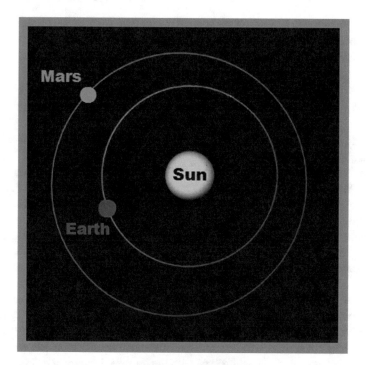

Fig. 10.5 Elliptical orbits of Earth and Mars around the Sun

Fig. 10.6 The distinct
distances of perihelion and
aphelion

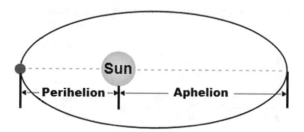

What was clearly 'visible' for Kepler from Tycho's data and what significantly differentiated Mars orbit from the original [simplistic and circular] version of the Copernican/heliocentric model - was that the Sun position is significantly and visibly displaced from the center by about 1/2 units in comparison to the approximate radius dimension of 5 units. But Mars orbit still appeared circular.

The **Perihelion** of any orbit of a celestial body orbiting the Sun is the point in the orbit where the body comes **nearest** to the Sun. It is the opposite of **Aphelion**, which is the point in the orbit where the celestial body is *farthest* from the Sun. The distinct distances of Perihelion and Aphelion of a generic orbit are depicted in Fig. 10.6. [Note: The Sun should appear further to the left for this highly eccentric ellipse].

From Fig. 10.6 we can ascertain that **Perihelion = A − C**, as well as that **Aphelion = A + C**.

Calculating Mars Perihelion and Aphelion in units of km, we get:

$$\text{Perihelion} = A - C = 227{,}936{,}637 - 21{,}292{,}093 = 206{,}644{,}544.$$
$$\text{Aphelion} = A + C = 227{,}936{,}637 + 21{,}292{,}093 = 249{,}228{,}730.$$

Clearly, Kepler could not have possibly detected the small discrepancy between the semi-major axis A and the semi-minor axis B for Mars, which is merely 227,936,637 − 226,939,986 = **996,651** km. As was mentioned earlier, Mars orbit appears nearly circular. Yet, Kepler was easily able to detect the discrepancy between the Perihelion and the Aphelion of Mars which is 249,228,730 − 206,644,544 = **42,584,187**. This significant and easier-to-detect discrepancy for the orbit of Mars regarding its deviation from circular motion with the Sun at the center of the circle is what prompted Kepler to arrive at his revolutionary idea of the ellipse. Perhaps it was quite a fortunate occurrence for science that Kepler's focus was placed on Mars, as opposed to focusing on planets Venus, Earth, Jupiter, and Saturn. This is so since all of these other planets—with the noticeable exception of Mercury—are of exceedingly low eccentricity and thus not only appearing totally circular-like for all practical purposes, but in addition their focal points are very much near the center, with only tiny displacement from it, and it would have been practically impossible for Kepler to detect their true elliptical nature even with the improved accuracy of Tycho's data. Another possibility is that Mars's unique discrepancy drew the attention of Kepler to it. Most likely Tycho did not have much data on Mercury due to its closeness to the Sun which makes observations quite difficult.

Planet	Eccentricity	Perihelion	Aphelion
Mercury	0.205631	46,375,340	70,310,999
Venus	0.006773	107,411,271	108,907,250
Earth	0.016709	146,605,913	152,589,828
Mars	0.093401	206,644,544	249,228,730
Jupiter	0.048495	740,524,420	816,624,857
Saturn	0.055509	1,352,544,268	1,514,498,923

The general view of the Solar System is then of nearly circular orbits for each and every planet; not appearing as ellipses in the least in spite of their elliptical nature; yet having some small displacements from their centers—except for Mercury and for Mars which show quite significant displacements from their centers. The eccentricity of Mercury is about double the eccentricity of Mars. Indeed Mercury is the most eccentric planet in the Solar System; therefore, with hindsight, analyzing Mercury's orbit would have been more beneficial or simply easier for Kepler than analyzing Mars' orbit (assuming he had enough data on Mercury).

Mercury's eccentricity is of the value 0.2056.
For Mercury, B is a bit less than A.

A = Semi-major axis is 57,909,050 km.

B = Semi-minor axis is 56,671,520 km.

C = Focal point = $\sqrt{(A^2 - B^2)}$ = 11,907,850 km.

$$e = \sqrt{(1 - B^2/A^2)} = \sqrt{(1 - 0.95772)} = \sqrt{(0.04228)} = 0.2056.$$

If we let 1 unit on the Cartesian Plane equals 11,458,630 km, then the dimensions of Mercury orbit converts into:

A = Semi-major axis is 5.05 units.

B = Semi-minor axis is 4.95 units.

C = Focal point = 1.04 units.

$$e = \sqrt{(1 - B^2/A^2)} = \sqrt{(1 - (4.95^2)/(5.05^2))}$$
$$e = \sqrt{(1 - 0.95773)} = \sqrt{(0.04227)} = 0.2056$$

Applying the above unit for distances, the elliptical orbit of Mercury with the Sun at one focal point is then depicted in Fig. 10.7, where the 'displacement' of the Sun from the supposed center is definitely more pronounced than that in the case of Mars.

Hence, even though Mercury's orbit is truly elliptical, it is not easy to visualize and to detect that its shape is not a circle, but with some effort it can be noticed. Figure 10.7 was drawn correctly, extending 10.11 units on the major (horizontal) axis, and 9.89 units on the minor (vertical) axis, and this discrepancy—however small—can almost be visualized here on the graph (with significant eye-effort).

The nearly circular orbit of the planets and their relatively low eccentricities is a property that renders the Solar System somewhat unique. By contrast, most known exoplanets [*planets discovered around stars within the Milky Way Galaxy but outside the Solar System*] have quite eccentric orbits. Approximately 40% of known exoplanets have eccentricities greater than 0.2, while about 14% have eccentricities greater than 0.4.

What is special about the Solar System that orbits of planets here are nearly circular, but elsewhere they are moderately or highly eccentric? One theory of planetary system evolution suggests that this is due to multi-planet interaction, predicting that multi-planet systems might have less eccentric orbits. The presence of many planets in a system induces planet-to-planet gravitational interactions and could indeed gradually and eventually dampen eccentricity. The stability of the Solar System is a subject of much inquiry and debate in astronomy. The planets' weak gravitational effects on one another can add up in unpredictable ways, accumulating over time. For this reason the Solar System is slowly evolving, and even the most

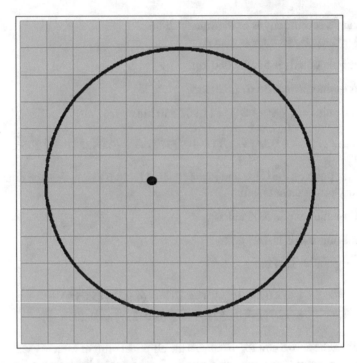

Fig. 10.7 Mercury's elliptical orbit with sun at a focal point

precise long-term models for the orbital motion of the Solar System are not valid over more than a few tens of millions of years.

Chapter 11
Kepler's 2nd Law—Differentiated Orbital Speeds

The motivation for Kepler's second law was a noticeable and decisive variation in Tycho's data on Mars regarding calculated speed. Apparently nobody before Kepler ever inquired about the speed of the planets; instead the focus was merely on location and timing. Kepler was the first to introduce this new focus on speed in astronomy. He noticed that as Mars was closest to the Sun around the Perihelion, it was moving the fastest; and as Mars was farthest away from the Sun around the Aphelion, it was moving the slowest. In other words, that the speed of Mars was positively correlated with its closeness to the Sun.

This observation must have seen natural to him, as it agreed with Kepler's hypothesis that the motive power radiated by the Sun drives the planets to move, and that this power weakens with distance, causing faster or slower motion as planets move closer or farther away from it.

Moreover, this differentiation between the two extreme poles of fastest and slowest at Perihelion and Aphelion was changing gradually, so that the speed seemed to be a direct function of the distance from the Sun. Kepler attempted several lines of attack to arrive at some sort of a pattern or a mathematical expression that would exactly account for the speed at every point—as implied by Tycho's data. Verifying any such relationship throughout the orbital cycle, however, required very extensive calculations. Kepler persevered, and after several failed attempts he had arrived at a somewhat indirect yet highly satisfactory statement that neatly accounted for the differentiation in speed. The statement is formulated in terms of geometry, without explicitly referring to speed, but it directly implies the observed variations in speed.

Kepler's second law:

A line segment joining a planet and the Sun sweeps out equal areas during equal intervals of time.

Figure 11.1 depicts two arbitrarily-selected areas of the elliptical orbit. The area on the left is near the Perihelion, where the planet speeds up. The area on the right is near the Aphelion, where the planet slows down. The time interval to pass through

A. E. Kossovsky, *The Birth of Science*, Springer Praxis Books, https://doi.org/10.1007/978-3-030-51744-1_11

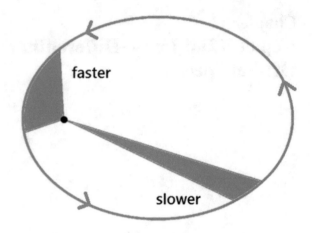

the long arc on the left [say 37 days] is equal to the time interval [also 37 days] to
pass through the short arc on the right. That means that the speed on the left near the
Sun is faster [more distance per given time], and that the speed on the right farther
away from the Sun is slower [less distance per given time]. If we construct any other
segment of an arc which also takes exactly 37 days to pass, then we can be assured
that calculated area there would be the same as in the case of the left area or the right
area seen in Fig. 11.1.

Alternatively, the conceptual framework of the second law can be stated more
clearly as follows: let us consider the passing of two distinct arcs, one near the
Perihelion and one near the Aphelion, with both taking the same amount of time
spent in travel, as in Fig. 11.1. The left area at Perihelion is akin to a fat but very
short man who weighs 77 k. The right area at Aphelion is akin to a skinny but very
tall man who also weighs exactly 77 k. Intuitively, regarding relative areas swept,
the arc on left near the Perihelion has the advantage that it moves faster; hence it is
wider and can potentially attain more area, yet it is disadvantaged geometrically in
sweeping less area because distance from the Sun to the arc is shorter. The arc on
right near the Aphelion has the disadvantage that it moves more slowly; hence it is
narrower and can potentially attain less area, yet it is advantaged geometrically in
sweeping more area because distance from the Sun to the arc is longer. Remarkably,
these two opposing factors cancel and offset each other exactly, so that resultant area
is neutral and indifferent to where the arc is located!

Surely this imaginary and ethereal 'pie in the sky' in the form of some invented
celestial area that Kepler dreams about and carefully draws many shapes of it on his
terrestrial wooden desk, does not really exists; it is just a mathematical abstraction.
Yet, it is precisely such pioneering and novel ideas that science desperately needed
at this juncture. Such strange abstractions would be followed in later generations by
even more mysterious and unseen ones, such as 'forces', 'vectors', 'electric flux',
'wave functions', and 'world lines'.

Kepler's success at arriving at his second law also depended on basing his entire work within the framework of the Copernican Sun-centered model. It would have been by far more difficult or rather nearly impossible for Kepler to make these grand discoveries had it all been based and formulated in terms of the complex and arcane Ptolemaic earth-centered model.

The extended line of research of Kepler culminated in the publishing of his second book **'Astronomia Nova'** ('A New Astronomy') in 1609, which includes his first and second laws of planetary motion.

Chapter 12
Kepler's 3rd Law—Harmony of the Planets

Following the discovery of his first and second laws and the publication of Astronomia Nova in 1609, Kepler spent the next ten years in futile effort searching for any further regularities or patterns in Tycho's data using a variety of imaginative but fruitless approaches. Finally, Kepler became intrigued by the idea that the orbits of the entire set of the six planets should satisfy some common mathematical relationship, comparable to the mathematical relations between harmonious musical tones discovered by Pythagoras. This line of thought led to his last remarkable breakthrough—the third law of planetary motion.

As Kepler later recalled: *"on the 8th of March in the year 1618, something marvelous appeared in my head"*. Although Kepler gives the precise date of this discovery, he does not give any details about how he arrived at this conclusion.

Let us narrate the philosophical background laid out by some of the ancient Greeks which helped in part to inspire Kepler in the discovery of the third law. Pythagoras was a Greek philosopher and mathematician who lived in the sixth century BC. He founded a philosophical school in southern Italy around 525 BC, which soon evolved into a cult and religious brotherhood. For the Pythagorean School, art, religion, music, science, and mathematics, were all intertwined. In addition they were strictly vegetarians, wore white clothes, believed in the transmigration or reincarnations of souls, and deemed the study of philosophy as a mean of spiritual purification.

A major principle or dogma of the Pythagoreans was the claim that whole numbers ruled the world, and that everything could be explained in terms of integers or fractions of integers. They held the mystical belief that reality is mathematical in nature, and they made important contributions to mathematics, but they also wrapped themselves in mystery, considering themselves the guardians of the secrets of mathematics from the impious and wicked world. The Pythagoreans outright worshipped their philosophical beliefs and mathematical work, going so far as to sacrifice an ox after

A. E. Kossovsky, *The Birth of Science*, Springer Praxis Books, https://doi.org/10.1007/978-3-030-51744-1_12

discovering the 47th Proposition of Euclid, namely that the lengths of a right triangle are as in $A^2 + B^2 = C^2$.

Pythagoreans interweaved rationalism and irrationalism more inseparably than did any other movement in ancient Greek thought. Their secretive ways of life caused the eventual loss of some of their important discoveries, and it also made it impossible to ascertain how much of their mathematical knowledge they owed to prior discoveries made in ancient Egypt and Mesopotamia (modern-day Iraq, western Iran, eastern Syria, and southeastern Turkey). One member—Hippassus of Metapontum—caused a scandal by presenting a rigorous mathematical proof that the square root of two [the diagonal of a square of unit one] is an irrational number that cannot be expressed as a ratio of two integers. This was a severe crisis for the Pythagoreans who dogmatically believed that only positive rational numbers could exist, and Hippassus' proof was unmasking their philosophical failure. They were so horrified by the idea of the square root of two that they vowed to keep the existence of irrational numbers an official secret of their sect.

Finally a general persecution of the Pythagoreans occurred. Many of the followers were killed or driven away. The Pythagorean meeting place was burned to the ground and Pythagoras was forced to flee with his remaining followers around 480 BC. It should be noted that in spite of their mysticism and superstitions, the contribution of the Pythagorean School to Western culture has been quite significant.

Pythagoras is said to be the first to make the connection between mathematics and music. According to legend, Pythagoras was once walking on the streets of Samos, when the sounds of blacksmiths' hammering the metal suddenly inspired him, prodding him to have a great insight. Some hammers produced *harmonious* sounds while others produced unpleasant *discordant* sounds. Pythagoras rushed into the shop to analyze mathematically the relative weights of the blacksmiths' hammers. To his great surprise the differences between harmonious and discordant sounds could be explained mathematically. He discovered that only when the weights of two hammers are as in whole-number ratio, such as say 1/2, 3/2, 9/8, 4/5, and so forth, striking them produced pleasant harmonious tones; otherwise they produced unpleasant discordant sounds. More generally, these ratios relate to the (vibration) frequency of the sounds produced—measured in Hertz (Hz).

Pythagoras' mathematization of music laid the foundations of music that today's many musicians are building upon. The achievement of the Pythagoreans in musical theory is not considered mystical, numerological, or controversial.

Pythagorean Tuning is a system of musical tuning in which the frequency ratios of all intervals are based on the ratio **3/2** plus some adjustments up or down using the factors 2/1, 1/2, and 2/3.

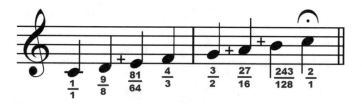

Fig. 12.1 Pythagorean tuning with multiples 3/2, 2/3, 1/2, 2/1

The **3/2** ratio—also known as the 'pure' perfect fifth—is chosen because it is one of the most consonant and easiest to tune by ear.

Figure 12.1 depicts an arrangement of repeated applications of the ratios 3/2, 2/3, 1/2, and 2/1. The successive fractions of whole numbers seen in Fig. 12.1 and shown in blue color are gotten by the cumulative multiplications of:

$$(1/1), (3/2)^2(1/2), (3/2)^2(1/2), (2/3)^5(2/1)^3, (3/2)^2(1/2), (3/2)^2(1/2),$$
$$(3/2)^2(1/2), (2/3)^5(2/1)^3$$

Pythagorean tuning can be said to be based on a stack of intervals called perfect fifths, each tuned in the ratio **3/2**—the next simplest ratio after 2/1. Wikipedia provides the table shown here which illustrates this—showing for each note in the basic octave: the conventional name of the interval from D (the base note), the formula to compute its frequency ratio, its size in cents, and the difference in cents (labeled ET-dif in the table) between its size and the size of the corresponding one in the equally tempered scale.

Note, Interval from D, Formula = Freq. Ratio, Size, ET-dif

Note, Interval from D	Formula = Freq. Ratio	Size	ET-dif
A♭ diminished 5th	$\left(\dfrac{2}{3}\right)^6 2^4 = \dfrac{1024}{729}$	588.27	−11.73
E♭ minor second	$\left(\dfrac{2}{3}\right)^5 2^3 = \dfrac{256}{243}$	90.22	−9.78
B♭ minor sixth	$\left(\dfrac{2}{3}\right)^4 2^3 = \dfrac{128}{81}$	792.18	−7.82
F minor third	$\left(\dfrac{2}{3}\right)^3 2^2 = \dfrac{32}{27}$	294.13	−5.87
C minor seventh	$\left(\dfrac{2}{3}\right)^2 2^2 = \dfrac{16}{9}$	996.09	−3.91
G perfect fourth	$\left(\dfrac{2}{3}\right) 2 = \dfrac{4}{3}$	498.04	−1.96
D unison	$\left(\dfrac{1}{1}\right) = \dfrac{1}{1}$	0.00	0.00
A perfect fifth	$\left(\dfrac{3}{2}\right) = \dfrac{3}{2}$	701.96	1.96
E major second	$\left(\dfrac{3}{2}\right)^2 \left(\dfrac{1}{2}\right) = \dfrac{9}{8}$	203.91	3.91
B major sixth	$\left(\dfrac{3}{2}\right)^3 \left(\dfrac{1}{2}\right) = \dfrac{27}{16}$	905.87	5.87
F♯ major third	$\left(\dfrac{3}{2}\right)^4 \left(\dfrac{1}{2}\right)^2 = \dfrac{81}{64}$	407.82	7.82
C♯ major seventh	$\left(\dfrac{3}{2}\right)^5 \left(\dfrac{1}{2}\right)^2 = \dfrac{243}{128}$	1109.78	9.78
G♯ augmented 4th	$\left(\dfrac{3}{2}\right)^6 \left(\dfrac{1}{2}\right)^3 = \dfrac{729}{512}$	611.73	11.73

Having once successfully established music as an exact science, Pythagoras then attempted to apply his newly found law of harmonic intervals to all the phenomena of Nature, even going so far as to supposedly demonstrate the harmonic relationship of the planets, constellations, and chemical elements to each other. A principal teaching of the Pythagorean School was that the world is in universal harmony, perceived through numbers.

Fig. 12.2 Kepler's third law
of planetary motion

$$\left(\begin{array}{c}\textbf{Time of one}\\\textbf{Full Revolution}\end{array}\right)^{2} = \text{K}\left(\begin{array}{c}\textbf{Distance}\\\textbf{from Sun}\end{array}\right)^{3}$$

In 1619 Kepler published his third book **'Harmonices Mundi'** ('The Harmony of the World') where he articulated what came to be known as his third law of planetary motion. The book is divided into five chapters. The first chapter is on regular polygons in geometry. The second chapter is on the congruence of figures in geometry. The third chapter is on the origin of harmonic proportions in music. The fourth chapter is on harmonic configurations in astrology. The fifth chapter is on the harmony of the motions of the planets in terms of his third law.

The third law of planetary motion—involving quantities raised to the powers of **3** and **2**—strongly reminds one of the ratio **3/2** of the harmonious Pythagorean Tuning as seem in Fig. 12.1. Indeed, Kepler's initial statement in his first draft of the third law was written exactly in terms of a **3/2** ratio, as shall be seen later. Incredibly, this exact ratio of the third law was probably a coincidental inspiration that Pythagoras bequeathed Kepler who lived two millennia after him!

Kepler's third law:

The square of the orbital period of a planet is equal to the cube of the distance of its orbit—measured as the semi-major axis. This equality—obtained via an adjusting constant depending on the units of length and time in use—is actually proportionality.

The third law depicted in Fig. 12.2 cannot be thought of as an independent statement about any individual planet (as were the first and second laws); rather it is a statement about the entire data set regarding all the planets. Kepler's third law is sometimes referred to as 'the law of harmonies' as it compares the orbital period and radius of orbit of any given planet to those of other planets, harmonizing them, and synchronizing them, so that all have the same relationship between time and distance. Unlike Kepler's first and second laws that describe the motion characteristics of a single planet, the third law makes a comparison between the motion characteristics of different planets. Here close attention must be paid to the time unit and length scale used in measuring astronomical data. Surely, the square of the orbital period (called T) should not normally equal to the cube of the distance from the Sun (called D), unless we calibrate our time and length units very carefully in such a way as to validate the relationship $T^2 = D^3$. Rather, the law incorporates a constant of proportionality called K, stating that for any given set of time and length units, the relationship $T^2 = KD^3$ holds true for all the planets. Equivalently, the third law states that the relationship $T^2/D^3 = K$ is valid for all planets, with K being the same fixed value no matter what planet is considered—so long as all the planets are measured consistently via an identical system of units and scales. Figure 12.3 depicts the application of the third law to Mars, Saturn, Earth, and Jupiter, while the application to planets Venus and Mercury are not shown here.

Fig. 12.3 The ratio form of
Kepler's third law

$$\frac{T_{Mars}^2}{D_{Mars}^3} = \frac{T_{Saturn}^2}{D_{Saturn}^3} = \frac{T_{Earth}^2}{D_{Earth}^3} = \frac{T_{Jupiter}^2}{D_{Jupiter}^3}$$

Fig. 12.4 Another form of
the third law for two planets

$$\left(\frac{T_{Jupiter}}{T_{Saturn}}\right)^2 = \left(\frac{D_{Jupiter}}{D_{Saturn}}\right)^3$$

Figure 12.4 depicts an alternative form of the third law involving only two planets. This is obtained by writing the ratio equality as in Fig. 12.3 for only two planets, followed by simple algebraic re-arrangements (cross-multiplication and powers for ratios). Arbitrary societal units or scales are canceled out within each ratio, resulting in a purely unit-less expression of the third law.

How is distance from the Sun to be measured? Should it be Perihelion or Aphelion? One cannot argue with Mother Nature; hence if she chooses one over the other then astronomical data would show that fact, and one should respect her preference and choice. As it happened, she has chosen wisely and fairly, being impartial to both by simply taking their average. Referring to Fig. 10.1, we recall that Perihelion = A − C; Aphelion = A + C; hence their average is (A − C + A + C)/2 = (2A)/2, or simply A, namely the semi-major axis.

Kepler is found to be the archetype and pioneer of the inductive scientific genius. He managed to unify in a most exacting manner all of the known planets orbiting the Sun in a way that had never been done before.

In Science, we often refer to two broad methods of reasoning, the deductive approach, and the inductive approach. Deductive reasoning works from the more general to the more specific. We begin with a broad theory or postulate, and then narrow that down into more specific hypotheses or corollaries that can be tested, and finally we collect observations to check the hypotheses or corollaries. This enables us to confirm or disprove our original theory or postulate.

Inductive reasoning works the other way, moving from specific observations to broader generalizations and theories. In inductive reasoning, we begin with specific observations and data, then attempt to detect patterns and regularities, and finally end up developing some general conclusion or theory.

Let us present the planetary data of that era that was available to Kepler for analysis, and see how he could have statistically or numerologically attempted to discern the inner pattern within the data leading to the discovery of the third law.

Kepler could not have utilized the Cartesian coordinate system [invented by René Descartes and named after him] in order to better visualize the data and plot a scatter plot (i.e. graph or chart) of the time of orbit versus distance from the Sun, or any variation of it, because the plane with x and y axes was simply not invented yet. Nowadays we are so used to reading charts and graphs that it is very hard for us to

Fig. 12.5 Discovery time of
the planets

Mercury	prehistoric time
Venus	prehistoric time
Earth	always!
Mars	prehistoric time
Jupiter	prehistoric time
Saturn	prehistoric time
Uranus	1781
Neptune	1846
Pluto	1930

imagine a time in history before the concept of a two-dimensional chart was even invented (by Descartes). We stare intensively with mixed emotions of fear and greed at charts of IBM stock price history going back perhaps 20 years in order to decide about the trend; should we buy it or should we sell it? Biologists and doctors stare intensively at charts of cholesterol level and a particular drug or some eating habits to learn about any possible correlation or relationship. But Kepler could not have utilized any such basic tools to get hints about possible relationships between the time of orbit and the distance from the Sun.

According to the Kepler scholar Liider Gabe, René Descartes visited the labs of Tycho Brahe in Prague and met Kepler for the first time in Regensburg in November 1620—two years *after* Kepler had published his third law. Moreover, René Descartes published his idea of the Cartesian coordinate system only in 1637, and it is almost certain that Descartes himself did not yet conceive of the idea when he met Kepler. Surely Kepler could not have used even the simplest tool of modern Statistical Data Analysis such as Linear Regression. The earliest form of regression analysis was the Method of Least Squares, which was published by Adrien-Marie Legendre in 1805, and by Carl Friedrich Gauss in 1809; then further expanded by Francis Galton around 1880 to describe data relating to biological phenomena.

Figure 12.5 depicts the discovery time of the nine major planets in the Solar System. Note: Pluto's planetary status is now in doubt.

So Uranus, Neptune, and Pluto were still unknown to Kepler.

He could only work with data on the other six planets.

Figure 12.6 depicts the distances from the Sun for the nine major planets in the Solar System. Distance is defined here as the semi-major axis, and measured in units of 1,000,000 km. Figure 12.7 depicts the time for one complete revolution in the orbits of the nine major planets in the Solar System.

For Newton, the crucial input from Kepler was the third law, as well as Kepler's vague reference to gravity in attributing to the Sun the power of a 'moving soul'. The first law does not carry any direct hint of classical mechanics. Even the second law—which implicitly reveals variations in planet's speed according to closeness to the Sun—alludes more to geometrical patterns than to mechanics. Moreover, the

Fig. 12.6 Distance from the
Sun

PLANET	Distance (10^6 km)
MERCURY	58
VENUS	108
EARTH	150
MARS	228
JUPITER	779
SATURN	1,434
URANUS	2,873
NEPTUN	4,495
PLUTO	5,906

Fig. 12.7 Time for one full
revolution

PLANET	Time (days)	Time (years)
MERCURY	88	0.2
VENUS	225	0.6
EARTH	365	1.0
MARS	687	1.9
JUPITER	4331	11.9
SATURN	10,747	29.4
URANUS	30,589	83.8
NEPTUNE	59,800	163.7
PLUTO	90,560	247.9

format of the third law, as a compact and simple relationship $T^2 = KD^3$ and its superb success in fitting astronomical data, surely have inspired Newton in stating the much more universal law of $F = MA$ and the law of universal gravitation $F = GM_1 M_2/R^2$.

In that sense, Kepler's last act in discovering the third law constitutes one of the few dramatic events in the long process of the birth of science. Let us then go through some hypothetical analytical steps that Kepler might have gone through on his way to the third law. Figure 12.8 depicts the data that was available to him. Surely Kepler must have actually written down these two variables together side by side in a single table as in Fig. 12.8 and stare at it intensively in order to analyze and understand the data, although perhaps with different units of time and distance.

It is incredible that this paltry data set—merely 2 variables of 6 data points each—could constitute the seed of modern physics!

Well, it takes a mathematician and a scientist such as Kepler to bring this seed into fruition via just 12 values.

PLANET	Time (days)	Distance (10^6 km)
MERCURY	88	58
VENUS	225	108
EARTH	365	150
MARS	687	228
JUPITER	4,331	779
SATURN	10,747	1,434

Fig. 12.8 Data available to Kepler to concoct the third law

The obvious conclusion that can be drawn from the available data as seen in Fig. 12.8 is that distance and time correlate positively. The farther a planet is from the sun, the more time it takes for it to complete one full revolution. Well, this is common sense, since the farther a planet is from the sun the wider is its circular or elliptical perimeter, and the more distance it has to travel, so surely it takes longer time to complete a revolution. Even for eccentric Mars and Mercury the shape of the orbit is nearly circular for all practical purposes; hence circumference is nearly $2\pi R$, or $2\pi D$ in our notations, so that the length of the orbital path is directly proportional to the distance from the Sun. Doubling the distance from the Sun for any given planet doubles its circumference, and so under the naïve assumption of equal speeds for all the planets, the time it takes for one complete revolution doubles as well. Hence perhaps an exact linear relationship between time and distance should be observed on the scatter plot if Kepler could have conceived of the possibility of drawing it.

Modern-day statisticians or scientists would most likely construct the scatter plot seen in Fig. 12.9 and would stare at it for clues. The 6 points are surely not linear, but might perhaps fit a rising parabola or some sort of exponential growth. The cause of science can be significantly advanced by understanding the message embedded within this data set, although modern-day statistical methods should be avoided, because their application would badly mislead here.

Luckily for science, Kepler did not know anything about Simple Linear Regression, Logistic regression, or Nonparametric Statistics, so he did not try to obtain some parameters for some standard statistical model and stop there. Instead he looked for deeper and more meaningful relationship; he looked for some more exact pattern.

As Kepler searched for the 'harmony' in the sky and for exact numerical relationship, he has almost certainly constructed the numerical table shown in Fig. 12.10 with its added column on the right showing the time-to-distance ratios *within* each planet—comparable yet different than his previous exploration of the ratios of the distances *between* planets and their fit into the ratios of spheres embedded between Platonic Solids.

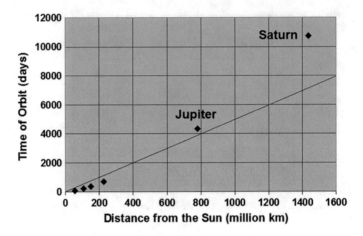

Fig. 12.9   Scatter plot of the data available to Kepler

Fig. 12.10   Time-to-distance
ratios for the six planets

PLANET	Time	Distance	Time / Distance
MERCURY	88	58	1.5
VENUS	225	108	2.1
EARTH	365	150	2.4
MARS	687	228	3.0
JUPITER	4,331	779	5.6
SATURN	10,747	1,434	7.5

Kepler's would have concluded from the available data as seen in Fig. 12.10 that the relationship between time and distance is not linear, and which is why time-to-distance ratios are not constant across planets. Nothing is steady, ratios are increasing, and time is growing much faster than distance. In other words: time gets to be much 'stronger' than distance the farther away we go from the Sun.

Surely Kepler could have guessed why [in a physical sense] the data gives the unmistaken message that time is 'stronger' here. The earlier simplistic argument regarding Fig. 12.8 about the need of the outer planets to traverse wider circumferences assumes that all the planets move at the same overall speed—regardless of their relative distance from the Sun. Kepler already guessed that this might not be so, and that the closer a planet is to the Sun the faster it moves, just as his second law implies! Hence if those planets farther away from the Sun move more slowly (i.e. overall average between Perihelion's speed and Aphelion's speed) and if those planets nearer the Sun move faster, then we might have two effects acting on orbit-time which neatly reinforce each other, namely, longer or shorter paths, as well as slower or faster overall speeds! Poor Saturn with its very long path has to travel a great deal of distance to complete just one revolution—while lucky Mercury with its very short path can easily complete its revolution. On top of this hurdle, poor Saturn

Fig. 12.11 Time-to-distance² ratios for the six planets

PLANET	Time	Distance²	Time / Distance²
MERCURY	88	3,352	$2.62 \cdot 10^{-2}$
VENUS	225	11,707	$1.92 \cdot 10^{-2}$
EARTH	365	22,380	$1.63 \cdot 10^{-2}$
MARS	687	51,938	$1.32 \cdot 10^{-2}$
JUPITER	4,331	606,218	$0.71 \cdot 10^{-2}$
SATURN	10,747	2,054,922	$0.52 \cdot 10^{-2}$

has to overcome another disadvantage of being very sluggish and much slower than speedy Mercury. Could this be the reason that Saturn takes about 30 years to perform what Mercury does in just 3 months?!

Kepler's next line of attack must have been the idea that since time is much 'stronger' than distance, then we need to counter this trend by 'strengthening' distance—thereby achieving more balance and fairness between time and distance. A neat way of helping distance to grow a bit faster is by squaring it. Hopefully, by squaring distance, its growth would be comparable to the growth of time, thus yielding a linear relationship between these two variables.

Consequently, Kepler might have constructed the numerical table seen in Fig. 12.11 showing the distance squared side by side with time, and with the added column on the right showing the time-to-distance-squared ratios.

Figure 12.11 is disappointing. This approach is not working. Nothing is steady. Ratios are decreasing now. We have greatly exaggerated with that squaring of the distance. We have permitted distance grow way too fast, and now it is easily overtaking time. Well, surely squares, cubes, and powers in general, are all associated with extremely powerful growth; therefore by not 'empowering' time in any way, while distance was squared, we have followed an erroneous and totally unbalanced approach.

Modern-day statisticians or scientists would most likely construct the time versus distance-squared scatter plot as in Fig. 12.12—so as to decide what to do next. The 6 points are surely not linear after this one-sided power transformation in favor of distance.

Okay then, so let's 'empower' both distance and time, to be fair and balanced to both. Surely squaring distance and squaring time wouldn't do, since the original data set indicated that time was 'stronger' than distance, and squaring both wouldn't change that disparity. What we need to do here is to empower distance to a slightly greater degree than the empowerment of time, so as to give distance a slight boost. Clearly, the next natural and obvious set of integers (exponents) compatible with the above discussion is 2 and 3, namely cubing distance and squaring time, since such data transformation helps distance a bit more than it helps time—while at the same time endows powers fairly to both of them.

Consequently, Kepler might have constructed the numerical table shown in Fig. 12.13 showing the distance cubed and the time squared, as well as the added column on the right showing $Time^2/Dist^3$ ratios.

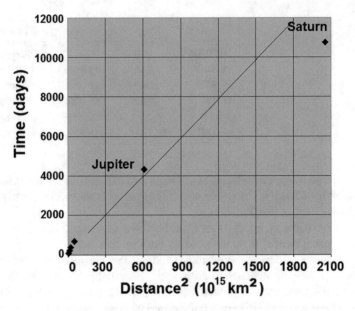

Fig. 12.12   Scatter plot of time versus distance-squared

PLANET	Time^2	Distance^3	Time^2 / Distance^3
MERCURY	7,744	194,105	$3.990 \cdot 10^{-2}$
VENUS	50,490	1,266,723	$3.986 \cdot 10^{-2}$
EARTH	133,371	3,348,072	$3.984 \cdot 10^{-2}$
MARS	471,969	11,836,764	$3.987 \cdot 10^{-2}$
JUPITER	18,757,561	472,001,304	$3.974 \cdot 10^{-2}$
SATURN	115,498,009	2,945,731,045	$3.921 \cdot 10^{-2}$

Fig. 12.13   Time^2-to-distance^3 ratios for the six planets

Eureka! It works! Now ratios are extremely steady across planets! This is the third law! Constant K ≈ 0.03974, for a system of units with day for time, and a million kilometers for length.

Modern-day statisticians or scientists would most likely construct the scatter plot seen in Fig. 12.14, and would be quite delighted to find the 'fitting curve' joining the 6 points to be almost perfectly linear. They would strongly validate this particular power transformation of the data, and enthusiastically confirm and certify the third law of planetary motion.

But the supposed process of discovery outlined above may not have been the way Kepler actually worked it out. Deprived of Statistical Data Analysis such as Linear Regression, and unable even to plot a scatter plot without knowing about the

Fig. 12.14 Scatter plot of
time squared versus distance
cubed

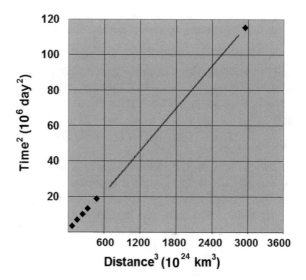

Cartesian coordinate system, Kepler might have been decisively aided by the recent mathematical invention of Logarithms just a few years before his discovery of the third law.

Let us briefly explain what **Logarithms** are for readers who are not familiar with the concept. We first need to understand the concept of 'power', which is simply repeated multiplications of a fixed base number called B, multiplied by itself P times. The notation B^P reads 'B to the power of P' and it means 'B multiplied by itself P times. If B = 2 and if P = 3, then 2^3 reads '2 to the power of 3' and it means '2 multiplied by itself 3 times', namely 2 * 2 * 2 = 8. Logarithm is simply a riddle or a puzzle relating to powers, such as $LOG_{10}(100)$ which reads 'the logarithm base 10 of 100', and which asks 'what power do I need to raise 10 to—in order to obtain 100?' and the obvious answer is 2, because $10^2 = 100$, hence $LOG_{10}(100) = 2$. In the same vein, $LOG_2(8) = 3$, namely the power you need to raise 2 to—in order to obtain 8, and which is simply 3, because $2^3 = 8$. Other examples are $LOG_5(25) = 2$, and $LOG_{10}(1,000,000) = 6$. Powers and logarithms do not necessarily have to be integers, and extrapolations to fractional and real values are also possible, such as in $LOG_5(21) = 1.8916$, and this is so because $5^{1.8916} = 21$. Surely we cannot multiply 5 by itself 'almost twice' or 'a bit less than twice', but allowing fractional exponents is indeed possible and in some cases necessary.

Let us quote Kepler's own account of how the third law came to be as outlined in Max Caspar's book 'Kepler' page 286:

> On the 8th of March of this year 1618, if exact information about the time is desired, it appeared in my head. But I was unlucky when I inserted it into the calculation, and rejected it as false. Finally, on May 15, it came again and with a new onset conquered the darkness of my mind, whereat there followed such an excellent agreement between my seventeen years of work at the Tychonic observations and my present deliberation that I at first believed that I had dreamed and assume the sought for in the supporting proofs. But it is entirely certain

and exact that the proportion between the periodic times of any two planets is precisely one and a half times the proportion of the mean distances.

As shall be demonstrated soon, the word 'proportion' in Kepler's last phrase basically means 'logarithm of the proportion', namely $LOG(D_1/D_2)$ or $LOG(T_1/T_2)$—instead of the proportion itself. Surely the 'ratio' of two quantities can often be termed or interpreted as 'proportion'. Hence Kepler is asserting here that:

$$LOG(T_1/T_2) = (3/2) * LOG(D_1/D_2)$$

Interestingly then, it appears that Kepler initially described his third law in terms of **3/2** ratio (i.e. **1.5** ratio) of the logarithm of proportions, rather than in the more familiar form of squared periods and cubed distances!

Let us derive the 3/2 ratio of logarithms in Kepler's initial insight from the familiar power-styled form of the law he used later as in Fig. 12.3 for multiple planets A and B:

$$T_A^2/D_A^3 = T_B^2/D_B^3$$
$$T_A^2/T_B^2 = D_A^3/D_B^3$$
$$(T_A/T_B)^2 = (D_A/D_B)^3$$

Which is actually the third law as in the format of Fig. 12.4.

Taking logarithm on both sides of the equation, and using the logarithm rule $LOG_B(N^P) = P*LOG_B(N)$, we get:

$$Log[(T_A/T_B)^2] = Log[(D_A/D_B)^3]$$
$$2 * Log[T_A/T_B] = 3 * Log[D_A/D_B]$$
$$Log[T_A/T_B] = (3/2) * Log[D_A/D_B]$$

Hence in a sense, as seen here, the third law is very much **logarithmic harmony** as well!

John Napier of Scotland published Mirifici Logarithmorum Canonis Descripto in 1614, regarding the novel invention of logarithms, and Kepler who soon became aware of it, may have spontaneously applied this fine new mathematical tool for his own collection of astronomical data. According to Kevin Brown, (Reflections on Relativity, 2016) Kepler was immediately enthusiastic about logarithms when he read Napier's work in 1616. In addition, around that time, Joost Bürgi of Switzerland published his work on logarithms, and Kepler was well-aware of Bürgi's work. In 1604 Bürgi entered the service of Emperor Rudolf II in Prague, and there he befriended Kepler and worked closely with him at the royal court. Bürgi's method is a bit different from that of Napier and was clearly invented independently. Kepler wrote about Bürgi's logarithms in the introduction to his Rudolphine Tables in 1627: "as aids to calculation Bürgi was led to these very logarithms many years before

Fig. 12.15 Log-transformed
data of time and distance

PLANET	Distance	Log(D)	Time	Log(T)
MERCURY	58	1.76	88	1.94
VENUS	108	2.03	225	2.35
EARTH	150	2.17	365	2.56
MARS	228	2.36	687	2.84
JUPITER	779	2.89	4331	3.64
SATURN	1434	3.16	10747	4.03

Napier's system appeared; but being a sluggish man, and very uncommunicative, instead of rearing up his child for the public benefit - he deserted it (i.e. hid it) at birth."

Therefore, although historians cannot be sure, it is plausible that Kepler used logarithms to discover the third law. It is interesting to note that what was primarily a purely mathematical innovation, namely logarithms, might have led directly to the formulation of the third law, and thus significantly aided in the discovery of modern physics. It should be noted that the concept of logarithm is primarily a mathematical one, but it also has a secondary aspect to it, namely being of immense aid in arithmetical calculations. Indeed, long before the electronic calculator and the computer were invented, books of logarithmic tables were published regularly for the purpose of easing the burden of arithmetical calculations for scientists, engineers, and other quantitative professionals. As shall be discussed at the end of this chapter, logarithm tables greatly helped Kepler in his tedious arithmetical calculations beginning in 1616.

Let us now focus on the formulation of Kepler's third law on a planet-by-planet basis as in Fig. 12.2, and apply logarithms to both sides of the equation, followed by the use of the logarithm rule $LOG_B(N*M) = LOG_B(N) + LOG_B(M)$, thus we get:

$$T^2 = K * D^3$$
$$Log(T^2) = Log(K * D^3)$$
$$2 * Log(T) = Log(K) + 3 * Log(D)$$
$$Log(T) = Log(K)/2 + (3/2) * Log(D)$$

Hence, from modern-day mathematical and statistical perspective, Kepler's third law asserts that the log-transformed data has a simple linear relationship with a slope value of 1.5 when Log(T) is expressed as a function of Log(D).

Let us verify this empirically using decimal base 10 logarithm.

Figure 12.15 depicts the table that Kepler might have arranged for himself in order to arrive at the third law. The units of day for time and one million kilometers for distance are used.

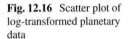

Fig. 12.16 Scatter plot of
log-transformed planetary
data

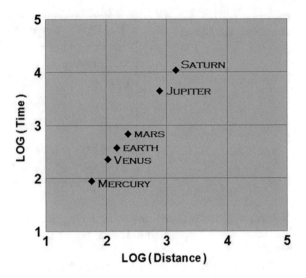

Figure 12.16 depicts the scatter plot of the log-transformed data of Fig. 12.15. The linearity of the plot can be clearly perceived.

Surely Kepler did not have the privilege of staring at the scatter plot of Fig. 12.16; moreover, the very idea of a slope or derivative was quite alien to him; instead Kepler focused on 'logarithm of the proportion' between various pairs of planets to find out the consistent 3/2 ratio. The 3/2 ratio is indeed the slope in Fig. 12.16 for any two points (two planets). Let us write his log-formulated law above twice, for any two generic planets A and B, thus we get:

$$Log(T_A) = Log(K)/2 + (3/2) * Log(D_A)$$
$$Log(T_B) = Log(K)/2 + (3/2) * Log(D_B)$$

Subtracting the 2nd equation of B from the 1st of A, we get:

$$Log(T_A) - Log(T_B) = (3/2) * \left[Log(D_A) - Log(D_B) \right]$$
$$\frac{Log(T_A) - Log(T_B)}{Log(D_A) - Log(D_B)} = \frac{\Delta Y}{\Delta X} = Slope = \mathbf{3/2}$$

Figure 12.17 depicts four actual such ratios of log differences (slopes) between arbitrarily selected pairs of planets (points).

It is probable that Kepler himself had arranged the same type of table as in Fig. 12.17, albeit a bit larger one showing several more planet-to-planet combinations. Surely the fit to 3/2 is superb!

Nowadays, the term **'logarithmic chart'** or **'logarithmic scale'** stands for plotting the logarithms of two variables, instead of the usual practice of plotting the original values themselves. This is done simply by first converting all the data values into their

Fig. 12.17 Some selected
ratios of log differences

PLANET	Log(D)	Log(T)	Ratio of Diff.
MERCURY	1.76	1.94	**1.497**
SATURN	3.16	4.03	
MERCURY	1.76	1.94	**1.500**
MARS	2.36	2.84	
EARTH	2.17	2.56	**1.501**
MARS	2.36	2.84	
VENUS	2.03	2.35	**1.499**
JUPITER	2.89	3.64	

logarithmic equivalences, and then drawing a scatter plot for these two newly created variables. A logarithm scale is one where intervals between the original data values of 1–10, 10–100, 100–1000, 1000–10,000, and so forth, all have the same width on the X-Log and Y-Log axes, say 1 inch, or perhaps 1 cm, even though actually these intervals are, according to the original numbers, expanding and widening constantly by a factor of 10.

Kepler's discovery of the third law in 1618 was based solely on the six known planets of his era. Uranus, the seventh planet from the Sun, was discovered by Friedrich Wilhelm Herschel in 1781, many years after Kepler's death. This was the first planet to be discovered since antiquity, and the overjoyed astronomers all around Europe and the world over cherished this discovery a great deal. Uranus' distance from the Sun and the time of its orbital period perfectly fit Kepler's third law, and this strongly validated Kepler's work. In 1801 the asteroid Ceres was discovered and this also strongly confirmed Kepler's third law. Ceres is the largest object in the Asteroid Belt orbiting between Mars and Jupiter. Later, Neptune was discovered in 1846, and the icy dwarf planet Pluto was discovered in 1930, both of which also strongly confirmed the third law.

Figure 12.18 depicts Kepler's postmortem decisive triumph and success regarding his third law for the 10 planets or planet-like bodies in the Solar System using the logarithmic scale. Figure 12.18 uses the units of day for time and one million kilometers for distance. The 3/2 ratio is indeed the slope in Fig. 12.18, and this obvious fact can be easily calculated. The graph shows the more comprehensive and nearly complete modern planetary data set for the Solar System falling on a nearly perfect straight line when converted to its logarithmic equivalences, according to Kepler's third law.

Logarithmic scale would yield a linear result on the scatter plot if X and Y relate via some 'power law' such as $Y^M = K * X^N$ or via some 'exponential law' Such as $Y = Q * B^X$.

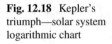

Fig. 12.18 Kepler's triumph—solar system logarithmic chart

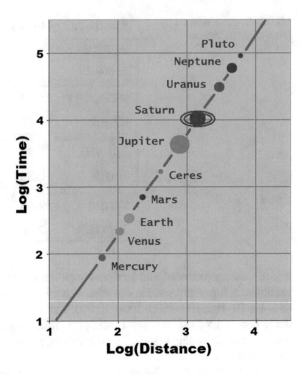

Even though power transformations of data and log transformations of data are radically different things, both yield straight line for data structured according to the power law $Y^M = K * X^N$ as is the case for the distances and periods of the planets as per Kepler's third law—assuming though the particular power transformation of the form $X^N \rightarrow X_{NEW}$ and $Y^M \rightarrow Y_{NEW}$ and the log transformation of $Log(X) \rightarrow X_{NEW}$ and $Log(Y) \rightarrow Y_{NEW}$. In addition, the slope of the scatter plot $Log(Y)$ versus $Log(X)$ will be N/M, while the slope of the scatter plot Y^M versus X^N will be K.

A modern-day statistician might have been able to get a hint of Kepler's third law if some kind of power law relationship was suspected, using linear regression style of approach and charts via the log-transformed or power-transformed data. But the success of such hypothetical statistician would depend heavily on letting go of the random approach altogether, letting go of parameters, and letting go of any correlation or regression models, and then look instead for an exact relationship between the distance to the Sun and the time period for one full revolution.

Napier and Bürgi came on the scene with the invention of the logarithm when Kepler was deeply immersed in mind-numbing, tedious arithmetical calculations. He had already been filling hundreds of folio pages with lengthily arithmetic operations, in his construction of the orbits of the planets from the observation of Tycho's data in the previous years. To Kepler, this discovery was a gift from heaven, for logarithms reduced considerably the time he had to spend just doing arithmetic calculations, a task which he greatly detested. Unfortunately for Kepler, all his previous work on

No.	Log.	No.	Log.	No.	Log.
1	0.000000	21	1.322219	41	1.612784
2	0.301030	22	1.342423	42	1.623249
3	0.477121	23	1.361728	43	1.633468
4	0.602060	24	1.380211	44	1.643453
5	0.698970	25	1.397940	45	1.653213
6	0.778151	26	1.414973	46	1.662758
7	0.845098	27	1.431364	47	1.672098
8	0.903090	28	1.447158	48	1.681241
9	0.954243	29	1.462398	49	1.690196
10	1.000000	30	1.477121	50	1.698970
11	1.041393	31	1.491362	51	1.707570
12	1.079181	32	1.505150	52	1.716003
13	1.113943	33	1.518514	53	1.724276
14	1.146128	34	1.531479	54	1.732394
15	1.176091	35	1.544068	55	1.740363
16	1.204120	36	1.556303	56	1.748188
17	1.230449	37	1.568202	57	1.755875
18	1.255273	38	1.579784	58	1.763428
19	1.278754	39	1.591065	59	1.770852
20	1.301030	40	1.602060	60	1.778151

Fig. 12.19 Logarithmic table in old manual books

the first and the second planetary laws in the years before 1616 had to be done the hard way without the benefit of algorithms.

Figure 12.19 is an example of **logarithmic table** published in logarithmic books for over three centuries since Napier and Bürgi era and until the advent of the calculator and the computer. These arithmetic manuals were published in order to aid in the calculations of difficult multiplications and divisions.

The logarithmic rules $LOG_{10}(N \times M) = LOG_{10}(N) + LOG_{10}(M)$ and $LOG_{10}(N/M) = LOG_{10}(N) - LOG_{10}(M)$ are the keys to the application of logarithms in arithmetical calculations, since they allow the conversion of multiplication into simple addition, and the conversion of division into simple subtraction.

As an example, the division of $57/19 = 3$ can be calculated by considering the logarithm of this division, namely $LOG_{10}(57/19)$ and then using the logarithmic table in Fig. 12.19 to convert this division into a simple subtraction of $LOG_{10}(57) - LOG_{10}(19) = 1.755875 - 1.278754 = 0.477121$. This logarithmic value of 0.477121 in turns points to the value 3 since it is the logarithmic equivalent of 3, as can be seen in Fig. 12.19. It is by far easier and much quicker to add or to subtract two numbers using a piece of paper and a pencil than to multiply or to divide numbers.

The table in Fig. 12.19 is actually not the most typical logarithmic table published in the old days, since it has a great deal of unnecessary duplication and repetitions. For example, $LOG_{10}(5)$ which is 0.698970 and $LOG_{10}(50)$ which is 1.698970 are redundant pieces of information. This is so since $LOG_{10}(50) = LOG_{10}(10 \times 5) =$

$LOG_{10}(10) + LOG_{10}(5) = 1.000000 + 0.698970 = 1.698970$, so that there is really no need to specify the logarithm of 50 as it can be easily deduced from the logarithm of 5.

The table in Fig. 12.19 also suffers from lack of numerical refinement, as it does not evaluate non-integral numbers with fractional values such as 2.378, 5.491, 8.673, and so forth. The typical logarithmic tables published in old manual books provide logarithmic values for the very long and refined set of numbers from say 1.00000 to 9.99999, in small incremental steps of 0.00001, as opposed to the redundant and arbitrary short set of integers from 1 to 60 of Fig. 12.19.

From this refined set of 1.00000 to 9.99999 numbers (pointing to logarithmic values between 0 and 1 exclusively, namely only fractions) one can also deduce the logarithm of any number below 1 or any number above 10. For example, to find out the logarithm of 8537, namely $LOG_{10}(8537)$, we manipulate the number and break it into two factors. One factor is 8537/1000 (i.e. moving the decimal point to the left 3 times), yielding 8.537, and which nicely lies within the interval (1, 10) of the logarithm table. The other factor is 1000 (i.e. moving the decimal point back to the right 3 times). This allows us to write $LOG_{10}(8537)$ as $LOG_{10}(1000 \times 8.537)$, which is easily evaluated as $LOG_{10}(1000) + LOG_{10}(8.537)$ or as $(3) + (0.931305)$ [obtained by reading the table], and which is simply 3.931305.

Hence, assuming the more refined logarithmic table from 1.00000 to 9.99999, a better arithmetical example can be given with the more challenging division Kepler might have had to perform, say the division of 567,525/329.

$$LOG_{10}(567,525/329)$$
$$LOG_{10}(567,525) - LOG_{10}(329)$$
$$LOG_{10}(100,000 \times 5.67525) - LOG_{10}(100 \times 3.29)$$
$$LOG_{10}(100,000) + LOG_{10}(5.67525) - LOG_{10}(100) - LOG_{10}(3.29)$$

Let us evaluate these 4 terms.
Surely $LOG_{10}(100,000) = 5$, and $LOG_{10}(100) = 2$.
Looking up the refined logarithmic table, we notice that:

$$LOG_{10}(5.67525) = 0.753985$$
$$LOG_{10}(3.29) = 0.517196$$

Hence the above 4 terms are reduced to:

$$5 + 0.753985 - 2 - 0.517196$$
$$3 + 0.236789$$
$$3.236789$$

Hence, Kepler could rapidly and with ease figure out that the logarithm of 567,525/329 is 3.236789. And now he only needs to figure out which number is such that its logarithm is indeed 3.236789. Focusing only on the fractional part 0.236789, and looking up the refined logarithmic table again, Kepler could quickly notice that $LOG_{10}(1.725) = 0.236789$.

Kepler can now cleverly add $LOG_{10}(1000)$ to the left side of the above equality, and add 3 to the right side of the above equality, because these two things are equal to each other, and because this way the right side becomes exactly 3.236789, thus he gets:

$$LOG_{10}(1000) + LOG_{10}(1.725) = 3 + 0.236789$$
$$LOG_{10}(1000 \times 1.725) = 3.236789$$
$$LOG_{10}(1725) = 3.236789$$

Hence Kepler can easily conclude that $567,525/329 = 1725$.

Interestingly, the widespread use of these manual books of logarithmic tables for calculations over the centuries eventually revealed a nearly universal pattern in the occurrences of numbers in the real world, and which is known nowadays as **Benford's Law**. These logarithmic books focusing on the refined set of 1.00000 to 9.99999 numbers, when in use for many years, demonstrate a differentiated degree in the physical wear and tear of the pages, and which is quite strong for low first digits [say between 1.00000 and 3.00000] and less so for high first digits [say between 7.00000 and 9.99999], hinting that numbers beginning with low digits occur quite frequently in real-life data, and that numbers beginning with high digits occur more rarely.

Benford's Law states that most real-life numbers in data sets start with low digits such as $\{1, 2, 3\}$ and only rarely with high digits such as $\{7, 8, 9\}$. According to the law, the expected proportions of the first digit on the left-most side of numbers in real-life data, having the nine possible digits $\{1, 2, 3, 4, 5, 6, 7, 8, 9\}$, is approximately as in $\{30.1\%, 17.6\%, 12.5\%, 9.7\%, 7.9\%, 6.7\%, 5.8\%, 5.1\%, 4.6\%\}$. Hence, about 30.1% of all numbers in real-life data begin with digit 1, while only about 4.6% of numbers begin with digit 9.

This counterintuitive and very surprising fact about real-life numbers in scientific data and in practically any other factual data is supported by solid empirical confirmation when data is tested for its digital configuration. The phenomenon has also earned several mathematical and statistical theoretical explanations, collaborating this consistent and ubiquitous empirical finding.

The explicit formula provided by Benford's Law, expressing the probability of the occurrence of the first digit on the left-most side of numbers in real-life and scientific data is $LOG_{10}(1 + 1/digit)$. According to this author—who has studied and investigated this intriguing phenomenon for well over a decade—the generic origin of this law is not about digits or numbers but about quantities, and that nearly all real-life histograms fall to the right overall, having positive skewness, meaning that the small is numerous and that the big is rare in the world. This insight eventually leads to the more general formula regarding quantities without any digital or numerical involvement, and it is termed **The General Law of Relative Quantities**.

Differentiated physical wear and tear of the pages in old logarithmic books for calculations is indeed quite noticeable, so that these books seemed to be more strained by use and quite worn in the first pages relating to first digits 1, 2, and 3, and progressively less so throughout the book for higher digits, culminating in the last pages relating to first digits 7, 8, and 9 which seemed to be in relatively excellent condition, as if they haven't been much in use, and all this indicates that numbers beginning with low digits such as {1, 2, 3} occur much more frequently in real-life data, while numbers beginning with high digits such as {7, 8, 9} occur more rarely.

Chapter 13
Implication of the Third Law to Orbital Speed

The implication of the third law to the speed of the planets is direct and nearly precise. As seen in Chaps. 9 and 10, the shape of the paths of the planets is for all practical purposes decisively circular; and that it is not of the oval-like highly eccentric elliptical curve. This is the case even for the slightly eccentric planets Mars and Mercury, as can be seen in Fig. 9.2 and Fig. 10.7. The circumference is then approximated to a very high degree as $2\pi R$, or simply $2\pi D$ using our notation of D for the distance from the Sun.

This first (and highly justified) circular approximation of the shape of the orbits is followed by another (and almost equally justified) approximation of a singular speed for any given planet via the averaging out of the variety of distinct speeds around the Perihelion, the Aphelion, and all the other middle locations. This average speed—in units of million kilometers per day—is calculated as the length of the entire circumference divided by the time it takes for one full revolution around the Sun.

The relationship $T^2 = KD^3$ between D and T according to the third law shall be applied in order to express T in terms of D. Taking square root on both sides of the equation we get:

$$T = \sqrt{(KD^3)} = \sqrt{(K)}\sqrt{(D^3)} = \sqrt{(K)}\sqrt{(D^2 D^1)} = \sqrt{K}\sqrt{(D^2)}\sqrt{(D^1)}$$
$$= \sqrt{K}D\sqrt{D}$$

Speed $\equiv$ [Distance]/[Time]

Speed $=$ [Circumference]/[Time]

Speed $= [2\pi D]/[\sqrt{KD}\sqrt{D}]$.

D cancels out in the numerator and the denominator, hence:

Speed $= [2\pi]/[\sqrt{K}\sqrt{D}]$

$$\text{Speed} = \frac{2\pi}{\sqrt{K}} * \frac{1}{\sqrt{D}}$$

Therefore the speed of a planet is inversely proportional to the square root of the distance. The bigger the value of D, the farther the planet is from the Sun,

© The Editor(s) (if applicable) and The Author(s), under exclusive license to Springer Nature Switzerland AG 2020
A. E. Kossovsky, *The Birth of Science*, Springer Praxis Books,
https://doi.org/10.1007/978-3-030-51744-1_13

Fig. 13.1 Planets' speed in
kilometer/Day

PLANET	SPEED
MERCURY	4,142,158
VENUS	3,030,068
EARTH	2,576,915
MARS	2,087,823
JUPITER	1,129,559
SATURN	832,467

Fig. 13.2 Bar chart of
planets' speed in
kilometer/day

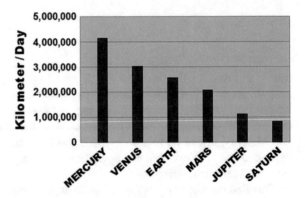

and the slower is the speed. This certainly hints at differentiated gravitational force
[emanating from the Sun] depending on distance. The planets near the sun must
possess a great deal of speed and inertia in order to combat the very strong gravitational pull exerted on them; while those farther away from the Sun with decreased
gravitational pull [due to great distance] move slower—without needing much speed
and inertia.

The approximated values of the constant K in time units of days and length units
of a million kilometers are shown in Fig. 12.13. The average value there of K =
0.03974 is chosen. Speed calculations in units of million-of-kilometers-per-day are
converted to kilometer-per-day units via the factor of 1,000,000.

Table in Fig. 13.1 depicts the speed of the 6 planets in units of kilometers per
day. We on Earth move on average a whopping distance of 2,576,915 km in one day!
Although to state it more precisely, faster speed occurs around Perihelion, and slower
speed occurs around Aphelion. The table was constructed by using the formula: Speed
$= 1{,}000{,}000 * [2\pi/\sqrt{0.03974}]/\sqrt{D}$.

The bar chart in Fig. 13.2 visually demonstrates the steady decline in the speed
of the planets as they are ordered from the ones closest to the Sun to the ones farther
away from it.

Chapter 14
The Third Law Is Independent of Planets' Mass

Kepler certainly understood at the time of his formulation of the third law that planets are not just points of light; rather, as Galileo's telescopic observations of the moon suggested, they are real physical entities just like our planet Earth, and perhaps they are made of sand, rocks, or ice, with hills and valleys, as the moon seemed to be. Galileo was Kepler's contemporary and they sometimes corresponded about their work and discoveries. Therefore Kepler must have reasoned that the planets are almost certainly of distinct sizes and distinct masses. Yet Kepler made no effort to account for or incorporate the planets' mass into his laws. But even if he had wished to incorporate mass, he couldn't, since nothing was known at the time about the masses of the planets.

Kepler might have found some comfort and assurance from Galileo's demonstration that projectile motion as well as vertical acceleration were independent of mass—such as in the legend of the two dropped rocks from the Tower of Pisa. Hence Kepler could have extrapolated Galileo's idea from rocks on Earth to those rocky planets in the sky around the Sun.

Kepler's three laws are indeed valid without any reference to mass. This follows from the well-known Equivalence Principle regarding the numerical equality between inertial mass and gravitational mass, and that therefore gravitational acceleration is independent of the mass of the body. Indeed, the Equivalence Principle was one of two seminal principles for Einstein in arriving at General Relativity; together with the principle that not only motion but also acceleration should be subjective and regarded as relative to the observers, and that only the laws of nature should be objective and invariant—confirmed by absolutely all observers even for those 'supposedly' accelerating frames of reference. Indeed, Kepler who knew of Galileo's discoveries, was vaguely aware of the Equivalence Principle and wrote about it in Astronomia Nova.

Let us briefly demonstrate how Newtonian Mechanics incorporates the Equivalence Principle for planetary motion.

Newton's 2nd law of motion $F = MA$

Universal Gravitation law $F = GM_1M_2/R^2$

© The Editor(s) (if applicable) and The Author(s), under exclusive license to Springer Nature Switzerland AG 2020
A. E. Kossovsky, *The Birth of Science*, Springer Praxis Books,
https://doi.org/10.1007/978-3-030-51744-1_14

Equating the right-hand sides of the two equations above for a small and light planet under the gravitational influence of the much bigger and more massive Sun (our star), we get:

$$(M_{PLANET})(A_{PLANET}) = (G)(M_{PLANET})(M_{STAR})/R^2$$

Assuming equality of gravitational and inertial masses, namely that the term (A_{PLANET}) on the left for inertia and forces is equal to the term (A_{PLANET}) on the right for gravitation, then (A_{PLANET}) cancels out on both sides of the equation, hence we get:

$$(A_{PLANET}) = (G)(M_{STAR}/R^2)$$

Namely that acceleration of the planet is independent of its mass.

The agreement of Kepler's three laws of planetary motion with the most precise data of the time, including subsequent telescopic data, was so spectacular that one might suppose that astronomers were immediate converts and highly enthusiastic about his discoveries. This was, however, far from the case.

Kepler's laws were not immediately accepted. Several major figures such as Galileo and René Descartes largely ignored Astronomia Nova and Harmonices Mundi. Galileo, who so courageously rebelled and dismantled many of the false Aristotelian's notions, still held on stubbornly to the Aristotelian's notion of the perfection of the circle for the motions of the heavenly bodies! And thus Galileo dismissed Kepler's notion of elliptical orbits! This fallacy might have been due to Galileo's admiration and deep trust in geometry in general, thus assuming that the simplest and most symmetric of all shapes, namely the circle, was the shape of the orbits. Galileo, his senior by just seven years, disappointed Kepler by not responding to his astronomical works even though the two corresponded on other scientific matters, and even though Galileo kindly recommended Kepler for a position as mathematician at the University of Padua when Galileo departed in 1611. Galileo must have been overwhelmed and too focused on his own extensive research to permit himself to spend much time analyzing Kepler's work and get distracted. In addition, Kepler's invocation of many mystical, esoteric, or metaphysical elements, such as the ethereal soul of the Earth, or the matric animal (moving soul) of the Sun led to some skepticism and caution. It was only after the ascendancy of Newton and his confirmation of Kepler's work that his three planetary laws were fully accepted.

Kepler has also significantly contributed to the foundation of modern optics. In 1604 Kepler published 'Astronomiae Pars Optica' ('The Optical Part of Astronomy') containing the inverse-square law governing the intensity of light, reflection by flat and curved mirrors, and principles of pinhole cameras, as well as the astronomical implications of optics such as parallax and the apparent sizes of heavenly bodies.

After hearing of Galileo's telescopic discoveries, Kepler started his own theoretical and experimental investigations of telescopic optics using a telescope borrowed from Duke Ernest of Cologne, Germany. It resulted in the publication of a manuscript in 1611 titled 'Dioptrice'. In it, Kepler set out the theoretical basis of double-convex converging lenses and double-concave diverging lenses—and how they are combined to produce a Galilean telescope. The manuscript also contains the concepts of real versus virtual images, upright versus inverted images, and the effects of focal length on magnification and reduction. He also described an improved telescope—now known as the astronomical or Keplerian telescope—in which two convex lenses can produce higher magnification than Galileo's combination of convex and concave lenses.

Kepler's later years were not happy ones as his life was increasingly disrupted by war. That was the Thirty Years' War, a bitter religious and dynastic conflict which pitted Protestants against Catholics, or rather that brutal king against another scheming king. The war began in Prague in 1618, and it then engulfed and brutally devastated many parts of Europe. In 1617 his mother Katharina was accused of witchcraft when a woman in a financial dispute with Kepler's family claimed that his mother made her sick with an evil brew. Witchcraft trials were relatively common in central Europe at that time, involving unimaginable and horrific brutality and cruelty against the innocents. Beginning in August 1620, his mother was imprisoned for fourteen months. She was released in October 1621, thanks in part to the extensive and successful legal defense drawn up by Kepler himself. Katharina was subjected to 'territio verbalis', a graphic description of the torture awaiting her as a witch, in a final attempt to make her 'confess'. The trial severely disrupted Kepler's life, yet in spite of it he managed to focus and continue his work and publications.

Chapter 15
Newly Invented Number System Facilitating Kepler's Planetary Calculations

The Roman Numeral system was in use in Europe for nearly 1800 years, far longer than the current Hindu-Arabic system has been in existence. The Roman Numeral system was developed around 500 BC, and it is partially of Greek origin as it descended from the ancient Etruscan Numerals—which itself is adapted from the Greek Attic symbols. As the Romans conquered much of Europe and the Mediterranean Basin, their numeral system spread and remained the primary manner for representing numbers.

The Roman system was rather inefficient and posed a serious obstacle to calculations. In addition, the lack of an effective system for utilizing fractions; the absence of the concept of zero; and no real or easy method for counting above several thousand and especially much larger values, rendered the Roman Numeral system even more cumbersome. Notwithstanding, it did not prevent ancient Rome's intellectuals and architects from building a great empire. Considerable mathematical skills were required to run a complex society and economy, and also to build vast monuments like the Colosseum and Constantine's Arch. In spite of the limitations of Roman Numerals, the existing archaeological record establishes that the Romans were able to overcome many of those obstacles with regard to the practicalities of construction. Roman roads and aqueducts remain as a testament to the engineering feats that the Romans were able to accomplish even with their flawed numerical system. Roman Numerals remain aesthetically important as seem in their widespread artistic use in architecture, clocks, and printing.

Addition and subtraction in Roman Numerals are not truly challenging and often are simple to perform, but multiplication and division are mostly difficult and time consuming, while fractional values further add to the complexity of use. Computation with the abacus (counting board tool) was quite common and, to some extent, overcame the difficulties of calculations with pen and paper. But the system certainly hindered progress in mathematics for many centuries. The transition from the Roman Numerals to the Hindu-Arabic numerals was surprisingly slow. It was not until around the sixteenth century that the Hindu-Arabic numerals became dominant in Europe.

© The Editor(s) (if applicable) and The Author(s), under exclusive license to Springer Nature Switzerland AG 2020
A. E. Kossovsky, *The Birth of Science*, Springer Praxis Books,
https://doi.org/10.1007/978-3-030-51744-1_15

A medieval mathematician, Leonardo of Pisa (1170–1240/50) better known as Fibonacci, published a book, 'Liber Abaci' ('The Book of the Abacus') in 1202. While working in North Africa, he had come upon the decimal system of notation used by Arab mathematicians, and he introduced it to Europe in this book. In the same publication he also described the famous sequence of numbers that bears his name—the Fibonacci Series. The Hindu-Arabic system most likely originated in India and came to Europe via the Arab world.

For Galileo the newly invented number system must have saved him a great deal of time and effort in his occasional calculations, but it wasn't crucial for his discoveries, since his work consisted mostly of conceptual and abstract thinking about the world, and designing and performing physical experiments, such as setting up pendulums, rock dropping, and projectile motion observations.

For young Newton aged 23, escaping from the plague at Cambridge, and resting at his little home village of Woolsthorpe, contemplating at leisure under the soft and cool shadow of a broad apple tree all that he had previously read about the work of Galileo and Kepler, there was no need for any calculations, numbers, easy arithmetic, or any efficient number system; his realm was purely in the abstract notions of gravity, motion, inertia, falling bodies, and so forth, and how to synthesize all these ideas into one coherent whole system of laws of motion.

Fortunately for Kepler, this essential civilizational background had been laid out and prepared for him just in time, as this new and incredibly efficient number system came about a generation before him, enabling him to perform his many essential arithmetical calculations regarding the timings, distances, and positions of the six known planets of his era.

Fortunately for Kepler as well, another essential civilizational background has been laid out for him just in time in the previous generation with Copernicus' formulation of the newly simplified and straightforward heliocentric model, essentially enabling Kepler to base all calculations and his new laws on this correct and easy-to-apply model of the Solar System.

Even with this indispensable utility of the newly invented efficient number system, Kepler still had to labor hard and long performing endless calculations, patiently and persistently. His calculations took up innumerable pages and hours upon hours of work, filling nearly 1000 large pages with dense arithmetical computations to obtain his famous laws of planetary motions.

Endowed with two qualities, which seemed incompatible with each other, a restless and fiery imagination as well as a persistent and stubborn intellect which the most tedious numerical calculations could not daunt, Kepler conjectured that the movements of the celestial bodies must be connected together by simple laws, or, to use his own expression, by harmonic laws. These laws he undertook to discover. A thousand fruitless attempts, errors of calculation inseparable from a colossal undertaking, did not prevent him a single instant from advancing resolutely toward the goal of which he imagined he had obtained a glimpse. Twenty-two years were employed by him in this investigation, and still he was not weary of it!

François Arago (1786–1853), a French mathematician, physicist, and astronomer.

Chapter 16
Numerical Example of the Complexity of Roman Arithmetic (Optional)

The letters {I, V, X, L, C, D, M} stand for the quantities {1, 5, 10, 50, 100, 500, 1000}. The numerical notation can be additive, as with 7 = VII and 11 = XI, as well as subtractive, as with 9 = IX and 4 = IV. Many clock designs mix these conventions, writing IIII for 4 and IX for 9. Figure 16.1 depicts the basic definitions and conversions of Roman Numerals into Hindu-Arabic numerals.

Caution should be exercised by present day professionals, engineers, and mathematicians when reading say MV, so as not to mistake its format as the standard notation for products such as $M * V$, namely as $1000 * 5 = 5000$, because here it's the addition signs '+' that is usually hiding invisibly between the Roman letters (or else the subtraction sign '−' at times when rank order is inverted), therefore the correct interpretation here is $MV = M + V = 1000 + 5 = 1005$.

Let us imagine Kepler having to multiply the time period for one complete revolution of say 86 days of Mercury, by say 41 million kilometers of its distance from the Sun. Let us consider the problem of calculating all this with Roman numerals, namely as LXXXVI (namely $50 + 10 + 10 + 10 + 5 + 1 = 86$) multiplied by XLI (namely $(50–10) + 1 = \mathbf{41}$). First we change to additive notation and consider LXXXVI by XXXXI. Then we take the product of each letter in the first number by each in the second. Formally this is called 'The Distributive Rule' in mathematics, namely that $Q * (A + B) = Q * A + Q * B$, and more generally that

$(G + H) * (A + B) = G * A + G * B + H * A + H * B$.

Thus we get:

$(LXXXVI) * (XXXXI) =$

$L * X + L * X + L * X + L * X + L * I$
$X * X + X * X + X * X + X * X + X * I$
$X * X + X * X + X * X + X * X + X * I$
$X * X + X * X + X * X + X * X + X * I$
$V * X + V * X + V * X + V * X + V * I$
$I * X + I * X + I * X + I * X + I * I$

© The Editor(s) (if applicable) and The Author(s), under exclusive license to Springer Nature Switzerland AG 2020
A. E. Kossovsky, *The Birth of Science*, Springer Praxis Books,
https://doi.org/10.1007/978-3-030-51744-1_16

Fig. 16.1   Roman numerals

1	=	I
2	=	II
3	=	III
4	=	IV
5	=	V
6	=	VI
7	=	VII
8	=	VIII
9	=	IX
10	=	X

I	1
V	5
X	10
L	50
C	100
D	500
M	1000

Using a Romanic Multiplication Table, which includes the following three results: $L * X = D$, $X * X = C$, $V *X = L$, we get:

$$D + D + D + D + L$$
$$C + C + C + C + X$$
$$C + C + C + C + X$$
$$C + C + C + C + X$$
$$L + L + L + L + V$$
$$X + X + X + X + I$$

Summarizing the sum of all the above values, we get:
We have 4 of $D = 4$ of $500 = 2000 = MM$
We have 5 of $L = 5$ of $50 = 250 = CCL$
We have 12 of $C = 12$ of $1000 = 1200 = MCC$
We have 7 of $X = 7$ of $10 = 70 = LXX$
We have 1 of $V = V$
We have 1 of $I = I$
Gathering them, consolidating them, and placing letters in their rank order, we get:
MMMCCCCLLXXVI
But $LL = C$, so this is simplified to MMMCCCCCXXVI, and since $CCCCC = D$, then this is further simplified to MMMDXXVI. In summary:

$$LXXXVI * XLI = MMMDXXVI$$

Let us now 'read' this Roman Numeral value in our numerical language, namely $1000 + 1000 + 1000 + 500 + 10 + 10 + 5 + 1$, which is simply 3526.

All this is quite cumbersome and fairly time consuming. Fortunately for Kepler, he would have been able to perform all this quickly and very easily—as we all learn to do nowadays at elementary schools, namely:

$$
\begin{array}{r}
86 \\
\mathbf{x}\quad 41 \\
\hline
86 \\
\mathbf{+}\quad 344\mathbf{0} \\
\hline
3526 \\
\end{array}
$$

Chapter 17
Galileo Galilei—The Father of Science

Galileo Galilei (1564–1642) was a central figure in the transition from natural philosophy to modern science and in the transformation of the early scientific Renaissance into a full Scientific Revolution. Few individuals have had as profound an impact on science as Galileo, whose groundbreaking inventions and discoveries earned him the title 'the father of science'. Galileo was an experimentalist who for the first time had the insight and talent to link theory with experiment. Galileo's innovative, experiment-driven approach to science all but disproved the Aristotelian physics and cosmology that had previously dominated the field in Europe. Figure 17.1 depicts a portrait of Galileo Galilei.

Galileo's lifework constituted a huge step towards the eventual separation of science from both philosophy and religion; a major development in human thought. Let us narrate briefly the broad historical background which shaped Galileo's early education, and which he later criticized and partly dismantled.

The adaptation of Christianity by the Roman Empire as the state religion around AD 323, and the subsequent fall of the empire around AD 476; were two significant events in European history. It was followed by a time of decline and loss in culture that lasted for centuries, an era called 'The Middle Ages' or 'The Dark Ages'. As Christianity spread in Roman territory, Greek philosophy and thoughts were suppressed. The most tragic loss was the burning of the Library of Alexandria in AD 415 by a mob of religious fanatics, just a few years after Paganism was made illegal in Alexandria by an edict of the Christian Emperor Theodosius I in AD 391. It was the biggest and the most significant library in the ancient world. The great thinkers of the age, scientists, mathematicians, and poets, from all civilizations came to Alexandria to study and exchange ideas at the library. As many as 700,000 scrolls filled its shelves. The burning of the Library of Alexandria, including the incalculable loss of ancient works, has become a symbol of the irretrievable loss of public knowledge. Yet, even during The Middle Ages, few books from Roman and Greek times survived and were still kept, mostly by monks in monasteries or other church centers around Europe— where some minimal form of general scholastic learning took place. In addition, the Arabic and Islamic world in the Middle East and North Africa served as a significant

© The Editor(s) (if applicable) and The Author(s), under exclusive license to Springer
Nature Switzerland AG 2020
A. E. Kossovsky, *The Birth of Science*, Springer Praxis Books,
https://doi.org/10.1007/978-3-030-51744-1_17

Fig. 17.1 Galileo Galilei

repository of the ancient Greek knowledge, safeguarding it, and then transmitting that knowledge back to Renaissance Europe when the continent suddenly showed an interest in what it had created long ago. From the ninth century until the fourteenth century, the Arabic civilization and caliphates adopted a positive approach to the sciences and gathered important Greek and mathematical books, safeguarding them, translating them into Arabic, and studying them. European translations of these precious books from Arabic to Latin beginning in the fourteenth century significantly contributed to the emergence of the Renaissance.

The Renaissance was a period from the fourteenth to the seventeenth century of cultural, artistic, and scientific awakening in European history. It started in Florence, Italy, in part due to the social and civic peculiarities of Florence at the time, and the patronage of its dominant family, the Medici, who supported the arts, architecture, and the sciences—and in particular supported Galileo in his scientific efforts. From Florence it then spread to other city-states in Italy, and later to the rest of Europe. Other factors leading to the Renaissance were the rediscovery of ancient scientific and mathematical texts via Arabic translations, as well as the Fall of Constantinople (the Byzantine Empire of Orthodox Christianity) in 1453 by an invading army of the Muslim Ottoman Empire, which generated a wave of expatriate Greek scholars bringing with them precious manuscripts in ancient Greek, many of which had fallen into obscurity in the West. Another factor leading to the Renaissance was the invention of book printing which popularized learning and allowed a faster propagation of new ideas. Johannes Gutenberg, in the German city of Mainz, developed European movable type printing technology around 1439, and in just over a decade, the European age of printing began.

Such was the fertile intellectual ground that Galileo was born into, right at the epicenter of the Renaissance, at Pisa Duchy of Florence, and he would then spend a lifetime nurturing and accelerating its progress.

Galileo showed a modern appreciation for the proper relationship or balance between (I) pure mathematics, (II) abstract theoretical or conceptual physics, and (III) experimental results. He along with Johannes Kepler were ones of the first modern thinkers to state that the laws of nature are mathematical.

Galileo was often willing to change his views in accordance with observation. In order to perform his experiments, Galileo had to set up standards of length and time, so that different measurements made in different locations and in different times could be compared in a reproducible fashion. This provided a reliable foundation upon which to perform his scientific work.

In his Metaphysics, Aristotle argued that since the things that mathematicians reason about are not material but rather abstract, therefore mathematics has to be excluded from the sciences and nature which are about the physical matter. Galileo fervently rebelled against this misguided idea. The Assayer publication in 1623 contains Galileo's famous statement that mathematics is the language of science, it reads: "Philosophy [i.e. physics] is written in this grand book—I mean the universe—which stands continually open to our gaze, but it cannot be understood unless one first learns to comprehend the language and interpret the characters in which it is written. It is written in the language of mathematics, and its characters are triangles, circles, and other geometrical figures, without which it is humanly impossible to understand a single word of it; without these, one is wandering around in a dark labyrinth."

Chapter 18
Galileo's Discoveries Regarding Falling Bodies

Galileo was interested in understanding how things moved and how they fell. What laws of motion governed them? Determining physical laws from experiment was a completely new undertaking; nobody has ever performed such a thing; but Galileo was intellectually adventurous enough to try. Galileo realized that out of all the observable motions in nature, free-fall motion is the key to the understanding of motion in general.

Galileo's publication of **Discourse and Mathematical Demonstrations Concerning Two New Sciences** in 1638 was his final book and a scientific testament covering much of his work in physics over the preceding forty years, and it directly paved the way for Newton's book Principia half a century later. Diverted for many years by his general interests in astronomy and his well-documented conflict with the Church, Galileo did not present his synthesis of his theory of motion until 1638. Aged, in his seventies, with failing eyesight, and in poor health, Galileo heroically wrote the Discourse while his house was regularly surrounded by Inquisition guards; under the threat of most dire consequences should he ever attempt to discuss the heliocentric hypothesis once again. One cannot but help admiring Galileo's tenacity. It was published in Holland, where the influence and power of the dreaded Inquisition was of less consequence. It was written just in time, because soon afterwards Galileo completely lost the sight of both his eyes, due to his past telescopic observations of the Sun. Tragically for Galileo, he would never actually see his published version of Two New Sciences; he would merely hold it in his ailing and tired hands.

Two New Sciences is rich with delightful investigations into the fundamental physical properties of the **strength of materials** and **motion**. Numerous geometrical diagrams are sprinkled throughout the book, as Geometry plays a central role in carrying out Galileo's analysis. The book is highly readable, even after more than three and a half centuries. A superb translation of the book into English was made by Henry Crew and Alfonso de Salvio in 1914, and now it is available online via the link:

http://galileoandeinstein.physics.virginia.edu/tns_draft/index.html

© The Editor(s) (if applicable) and The Author(s), under exclusive license to Springer Nature Switzerland AG 2020
A. E. Kossovsky, *The Birth of Science*, Springer Praxis Books,
https://doi.org/10.1007/978-3-030-51744-1_18

In Two New Sciences Galileo writes:

My purpose is to set forth a very new science dealing with a very ancient subject. There is, in nature, perhaps nothing older than motion, concerning which the books written by philosophers are neither few nor small; nevertheless, I have discovered some properties of it that are worth knowing that have not hitherto been either observed or demonstrated. Some superficial observations have been made, as for instance, that the natural motion of a heavy falling body is continuously accelerated; but to just what extent this acceleration occurs has not yet been announced. Other facts, not few in number or less worth knowing I have succeeded in proving; and, what I consider more important, there have been opened up to this vast and most excellent science, of which my work is merely the beginning, ways and means by which other minds more acute than mine will explore its remote corners.

Galileo chose a definition of uniform acceleration using what is sometimes known as the Rule of Parsimony: unless forced to do otherwise, assume the simplest possible hypothesis to explain natural events. Hence Galileo argued that as in the case of **constant speed** where we find no addition or increment simpler than that which repeats itself always in the same manner, where the intimate relationship between time and motion is such that uniformity of motion is defined by and conceived through equal times and equal spaces (thus we call a motion uniform when equal distances are traversed during equal time-intervals), so for **uniform acceleration** which we may define, in a similar manner, namely that during equal time intervals, a constant addition of speed takes place without complication.

Hence the definition of acceleration of motion which Galileo arrived at is stated as follows: A motion is said to be uniformly accelerating when it acquires during equal time intervals, equal increments of speed. [The symbol Δ indicates change in the value of something. For example, if a constantly moving car gains a distance of 160 km during the time interval between 7:45 am and 9:45 am, then its speed is calculated as (Δ Distance/Δ Time) = (160/2) = 80 km/h.]

$$\text{Constant Speed} = \frac{\Delta \text{ Distance}}{\Delta \text{ Time}} = \text{Constant}$$

$$\text{Constant Acceleration} = \frac{\Delta \text{ Speed}}{\Delta \text{ Time}} = \text{Constant}$$

There is no need to argue philosophically or test anything empirically if one merely defines things. Yet, to show its worth, Galileo had to take this construct one step further, namely stating that things in the real world actually fall that way.

Let us scrutinize Galileo's proposed definition. Is this the only possible way of defining uniform acceleration? Certainly not! Galileo's initial thoughts and analysis in 1604 had mistakenly led him to adopt an alternative [but misguided] definition of uniform acceleration where speed increases in proportion to the distance traveled as opposed to being in proportion to time elapsed. For example, it can be proposed that a dropped object accelerates downwards from an initial zero speed (rest), and that it gains additional speed of 3 m/s for each 1 m it falls, so that after falling 2 m down its speed is 6 m/s, and after falling 3 m down its speed is 9 m/s, and so forth. Is the falling object gaining a fixed and constant increment in speed per each second traveled or per each meter passed? Both definitions meet Galileo's requirement of simplicity. Both definitions seem to match our common sense idea of acceleration about equally well. How should Galileo choose between these two competing definitions? And surely there are other reasonable ways of defining the concept. By 1609 Galileo's theory had changed considerably, he had rejected the idea that speed increases with distance, and chose time instead as the basis of constant acceleration. It is noted that Galileo's journey of discovery of the laws of motion was lengthy, twisted, and anything but smooth!

In Two New Sciences Galileo writes:

> Although I can offer no rational objection to this or indeed to any other definition devised by any author whosoever, since all definitions are arbitrary, I may nevertheless without defense be allowed to doubt whether such a definition established in an abstract manner, corresponds to and describes that kind of accelerated motion which we meet in nature in the case of freely falling bodies.

Which definition would be more 'useful' or 'correct' in the description of nature? This is where experimentation becomes important. Galileo chose to define uniform acceleration as the motion in which the change of speed is proportional to elapsed time, and then strived to demonstrate that this matches the actual behavior of falling bodies. Galileo indeed made the right choice with his definition, yet he was not able to prove his case by direct or obvious means, due to difficulties in measuring bodies that fall so fast—lacking more modern and accurate clocks to measure time, and especially being unable to measure the difficult-to-measure variable of speed.

Let us devise an imaginary way for Galileo to demonstrate that his definition for uniform acceleration is indeed the correct one in describing the actual motions of falling bodies. The suggestion is that Galileo drops one heavy object from several different floors (i.e. of distinct heights) from the Leaning Tower of Pisa—with its top 57 m height. Galileo should then try to check whether the speed indeed increases in proportion to the time it takes to fall from the window to the ground. For each trial of each different floor Galileo should observe the time of fall and the speed just before the object strikes the ground.

But these two variables, namely time and speed, are very difficult to measure directly, even today. Moreover, since the average gravitational acceleration at sea level on Earth-denoted as g-is 9.80665 m/s^2, the entire time intervals of a fall from the top floor of the Leaning Tower of Pisa is just 3.41 s, and such time interval is shorter than Galileo could have measured accurately with the clocks available to him then.

But what would be exceedingly hard or rather impossible to measure here is the final speed (just before the object strikes the ground) which would involve measuring small intervals of motion-related time and length, a task which is by far more challenging than measuring large intervals of total time elapsed or total distance of displacement. In conclusion: direct test here was not possible for Galileo.

Galileo's inability to perform direct measurements to test his time-based constant acceleration hypothesis did not stop him. Ingeniously, he turned to mathematics to derive from his hypothesis another relationship not involving speed, but rather involving **total distance** and **total time**, which are by far easier to measure, and could be verified by experiments with available equipment.

An intuitive shorthand general reasoning follows:

Distance is proportional to **Speed*Time** [uniform or any motion]
Speed is proportional to **Time** [uniform acceleration]

Hence, by substituting 'Time' for 'Speed' in the first proportionality above we obtain:

Distance is proportional to **Time*Time** [uniform acceleration]

If all this can be expressed in an exact manner via a simple formula, then Galileo could potentially confirm his definition of acceleration directly by measuring only total fall time (squared) and total distance, while skipping the speed variable which is practically impossible to measure. Indeed, as shall be seen in the next (optional) chapter, Galileo arrived at the following concise formula:

Distance $=$ (Acceleration/2) $*$ Time^2
Distance $=$ (9.80665/2) $*$ Time^2
Distance $=$ (4.90333) $*$ Time^2

Chapter 19
Galileo's Elimination of Speed in the Calculation of Acceleration (Optional)

Consider an object that is initially at rest being dropped and thus experiencing free fall acceleration. Computing the total distance traveled is complicated by the fact that the speed is constantly increasing; hence the standard relationship (Speed) = (Distance)/(Time), or its equivalent relationship (Distance) = (Speed) * (Time) needs more refined and complex calculations.

As an example, if an object is dropped for 10 seconds, we might attempt to approximate the distance traveled for each second separately by using some sort of a 'middle speed' calculated at exactly the middle of each such time interval as the best approximation, namely calculating mid speeds at the times when the clock ticks 0.5, 1.5, 2.5, 3.5, 4.5, 5.5, 6.5, 7.5, 8.5, 9.5 seconds, and then sum up all these mini (Mid Speed) * (One Second) increments, to arrive approximately at the total distance traveled.

Since (Acceleration) = (9.80665) = (Speed)/(Time), according to Galileo's definition, it follows that (Speed) = (9.80665) * (Time), and more specifically at any time in the fall measured from the initial 0 time, (Current Speed) = (9.80665) * (Current Time), and therefore each increment of distance traveled for each second is approximately as in:

$$\text{Mini Distance} = (\text{Mid Speed}) * (\text{One Second})$$
$$\text{Mini Distance} = (\text{Current Speed}) * (\text{One Second})$$
$$\text{Mini Distance} = [(9.80665) * (\text{Current Time})] * (\text{One Second})$$

© The Editor(s) (if applicable) and The Author(s), under exclusive license to Springer Nature Switzerland AG 2020
A. E. Kossovsky, *The Birth of Science*, Springer Praxis Books,
https://doi.org/10.1007/978-3-030-51744-1_19

Fig. 19.1 Speed versus time
in constant acceleration

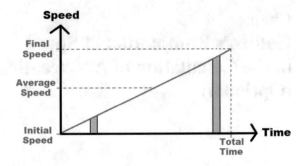

Thus the total Distance is:

$$[(9.80665) * (0.5)] * (1)+$$
$$[(9.80665) * (1.5)] * (1)+$$
$$[(9.80665) * (2.5)] * (1)+$$
$$[(9.80665) * (3.5)] * (1)+$$
$$[(9.80665) * (4.5)] * (1)+$$
$$[(9.80665) * (5.5)] * (1)+$$
$$[(9.80665) * (6.5)] * (1)+$$
$$[(9.80665) * (7.5)] * (1)+$$
$$[(9.80665) * (8.5)] * (1)+$$
$$[(9.80665) * (9.5)] * (1)$$

Calculating this sum, we obtain:
4.90 + 14.71 + 24.52 + 34.32 + 44.13 + 53.94 + 63.74 + 73.55 + 83.36 + 93.16 = **490.33** meters. Hence in 10 seconds the object has dropped approximately 490.33 meters.

Ultimately (in general) one should calculate more refined and smaller time intervals such as 0.1 s, or even 0.0001 s, in order to obtain more accurate figure for the total distance traveled. As it happened in this case, 1 s is indeed refined enough, sufficient, and exact, because of the linearity of the speed function, which allows things to be calculated geometrically.

But in general, for non-linear curves, this would require calculus integration, which would only be invented later by Newton in 1666 and Leibnitz in 1674. Let us first imitate Galileo's approach using modern day methods, and then describe Galileo's geometrically-based successful derivation.

Figure 19.1 depicts the graph of speed as a function of time in constant acceleration, for an object starting at rest with 0 initial speed.

The infinitesimal area of each 'infinitely thin' rectangle (i.e. trapezoid) is simply the Speed which is the height, multiplied by the infinitesimal Time increment which is the base (the width), and this area is equivalent to infinitesimal Distance via the relationship $D = S * T$.

Adding such infinitely many tiny rectangles yields the total area under the curve, and which here is simply the area of the 'right-bottom' triangle, having base T_{TOTAL} and height S_{FINAL}, namely:

$$D_T \equiv D_{TOTAL} = (1/2) * S_{FINAL} * T_{TOTAL} = (S_F T_T)/2 = (S_F/2) * T_T$$

$$\mathbf{D_T = (S_F/2) * T_T}$$

We have been using the geometric fact that the area of a right triangle is simply $(1/2) * (Width) * (Height)$. The above result $(S_F/2) * T_T$ could be interpreted as the average speed times the total time, and this is so since $S_{AVERAGE} = (S_{INITIAL} + S_{FINAL})/2 = (S_{FINAL} + 0)/2 = S_F/2$.

A more appropriate interpretation here would consider $S_F/2$ to be simply the midrange, standing exactly half way between the minimum of 0 and the maximum of S_F.

In addition, since Acceleration $= \Delta Speed/\Delta Time$, namely
Acceleration = (overall change in Speed)/(overall change in Time), we get:

$$Acceleration = (S_{FINAL} - S_{INITIAL})/(T_{TOTAL} - T_{INITIAL})$$
$$Acceleration = (S_{FINAL} - 0)/(T_{TOTAL} - 0)$$
$$Acceleration = A = S_{FINAL}/T_{TOTAL} = S_F/T_T$$

$$\mathbf{S_F = A * T_T}$$

Putting the earlier result of $D_T = (S_F/2) * T_T$ together with this new result of $S_F = A * T_T$, we obtain:

$$D_T = (S_F/2) * T_T$$
$$D_T = ((A * T_T)/2) * T_T$$
$$D_T = (A * T_T * T_T)/2$$
$$D_T = (A/2) * (T_T * T_T)$$
$$D_T = (A/2) * (T_T)^2$$

Namely that total distance (displacement) is proportional to total time squared (as well as to the value of the acceleration itself). Here, speed has been successfully eliminated completely from the equation! The alternative format of the above equation emphasizing the constant proportionality of D_T and T_T^2 is as follows:

$$\mathbf{D_T/T_T^2 = A/2}$$

$$\mathbf{DISTANCE_{TOTAL}/TIME_{TOTAL}^2 = (1/2) * Acceleration}$$

Fig. 19.2 Galileo's
remarkable insight about
acceleration

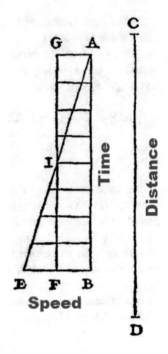

Surely one can [unnecessarily] apply Calculus and write D_T as:

$$D_T = \int_0^{T_T} \text{Speed}(t)dt = \int_0^{T_T} (At)dt$$

$$D_T = A\left(\frac{1}{2}\right)[T_T]^2 - A\left(\frac{1}{2}\right)[0]^2$$

$$D_T = \left(\frac{A}{2}\right)T_T^2$$

Let us now turn to Galileo, and the reasoning employed by him to arrive at the same conclusion. Galileo's detailed argument is based on geometrical consideration, where he draws a generic diagram for accelerated motion as seen in Fig. 19.2. The 'Time', 'Speed', and 'Distance' insertions in blue color are not of Galileo's writing; rather they are added by the author in order to aid the reader in following Galileo's line of thought.

Remarkably, by drawing Fig. 19.2, Galileo is—subliminally or almost sentiently—inventing the Cartesian Plane! And perhaps independently of its inventor René Descartes [who published his plane idea just a year earlier in 1637, although in a purely mathematical and totally non-physical context]. Galileo specifically mentioned that the drawing of the time axis AB is at right angle [perpendicular] to the speed axis EB! The positive direction of his horizontal axis is inverted though, pointing from right to left, as opposed to the Cartesian Plane which points from left to right for the positive direction. Moreover, it seems as if Galileo [in a totally subtle or subliminal manner] was performing a sort of Newtonian Calculus analysis many decades before it was invented, as he explicitly refers to the "sum of all the parallels", which presumingly assigns the term 'parallels' the meaning 'infinitesimal small rectangles'. To summarize, Galileo's argument is that the area under the rising line AE [i.e. area of triangle ABE] is equivalent to the area under the horizontal line GF [i.e. area of rectangle ABFG] with height $S_F/2$—being the middle value between 0 and S_F.

In Two New Sciences Galileo writes:

"Let us represent by the line AB the time in which the space CD is traversed by a body which starts from rest at C and is uniformly accelerated; let the final and highest value of the speed gained during the interval AB be represented by the line EB, drawn at right angles to AB; draw the line AE, then all lines drawn from equidistant points on AB and parallel to BE will represent the increasing values of the speed, beginning with the instant A. Let the point F bisect the line EB; draw FG parallel to BA, and GA parallel to FB, thus forming a parallelogram AGFB which will be equal in area to the triangle AEB, since the side GF bisects the side AE at the point I; for if the parallel lines in the triangle AEB are extended to GI, then the sum of all the parallels contained in the quadrilateral is equal to the sum of those contained in the triangle AEB; for those in the triangle IEF are equal to those contained in the triangle GIA, while those included in the trapezium AIFB are common. Since each and every instant of time in the time-interval AB has its corresponding point on the line AB, from which points parallels drawn in and limited by the triangle AEB represent the increasing values of the growing velocity, and since parallels contained within the rectangle represent the values of a speed which is not increasing, but constant, it appears, in like manner, that the momenta assumed by the moving body may also be represented, in the case of the accelerated motion, by the increasing parallels of the triangle AEB, and, in the case of the uniform motion, by the parallels of the rectangle GB. For, what the momenta may lack in the first part of the accelerated motion (the deficiency of the momenta being represented by the parallels of the triangle AGI) is made up by the momenta represented by the parallels of the triangle IEF. Hence it is clear that equal spaces will be traversed in equal times by two bodies, one of which, starting from rest, moves with a uniform acceleration, while the momentum of the other, moving with uniform speed, is one-half its maximum momentum under accelerated motion.

Chapter 20
Galileo's Experimental Confirmation of His Acceleration Theory

Yet, even with this successful elimination of the non-measureable speed variable from the equation relating distance and time for uniform acceleration, Galileo was still facing the severe challenge of measuring very short time intervals associated with free fall—as mentioned earlier.

So instead of directly testing his hypothesis, Galileo went one step further and invented an ingenious testing tool via balls rolling down an inclined plane which are of much lower rate of acceleration. By simply extrapolating the results from these rolling balls to freely falling bodies, he could indirectly validate his generic definition of acceleration. His hypothesis was that if a freely falling body has an acceleration that is constant, then a perfectly round ball rolling down a perfectly smooth inclined plane will also descend with constant acceleration, albeit at a significantly slower pace. But now, under such slower settings, the variables could be measured much more easily.

The use of an inclined plane allowed Galileo to dilute the force of gravity and slow the ball down considerably so that he could time it with a water clock, where he could measure and compare the weight of water that poured out before and after each experiment.

Here is how Galileo described his own experiments in Two New Sciences:

A piece of wooden molding or scantling, about 12 cubits long, half a cubit wide, and three finger—breadths thick, was taken; on its edge was cut a channel a little more than one finger in breadth; having made this groove very straight, smooth, and polished, and having lined it with parchment, also as smooth and polished as possible, we rolled along it a hard, smooth, and very round bronze ball. Having placed this board in a sloping position, by lifting one end some one or two cubits above the other, we rolled the ball, as I was just saying, along the channel, noting, in a manner presently to be described, the time required to make the descent. We repeated this experiment more than once in order to measure the time with an accuracy such that the deviation between two observations never exceeded one-tenth of a pulse beat. Having performed this operation and having assured ourselves of its reliability, we now rolled the ball only one-quarter of the length of the channel; and having measured the time of its descent, we found it precisely one-half of the former. Next we tried other distances, comparing the time for the whole length with that for the half, or with that for two-thirds, or three-fourths, or indeed for any fraction; in such experiments, repeated a full

© The Editor(s) (if applicable) and The Author(s), under exclusive license to Springer Nature Switzerland AG 2020
A. E. Kossovsky, *The Birth of Science*, Springer Praxis Books,
https://doi.org/10.1007/978-3-030-51744-1_20

Fig. 20.1 Inclined plane—Galileo museum of science, Florence

Fig. 20.2 Galileo's
proportionality for balls in
inclined planes

$$\frac{D_X}{T_X^2} = \frac{D_Y}{T_Y^2} = \frac{D_Z}{T_Z^2}$$

hundred times, we always found that the spaces traversed were to each other as the squares
of the times, and this was true for all inclinations of the plane, that is, of the channel along
which we rolled the ball.

The image in Fig. 20.1 reproduced by courtesy of Biografias y Vidas at www.
biografiasyvidas.com—depicts one such inclined plane from the Galileo Museum of
Science in Florence; although this particular plane is not the one that Galileo actually
used.

What Galileo found was that when a ball rolled down an inclined plane at a fixed
angle to the horizontal (i.e. fixed acceleration), the ratio of the total distance travelled
(i.e. the length of the incline) to the square of the corresponding time of travel was
always the same—regardless of the size (length) of the incline. For example, if D_X,
D_Y, D_Z represent distances (lengths from top to bottom) of three different inclined
planes with varying sizes (lengths), but of equal angle (and thus of equal acceleration),
and if T_X, T_Y, T_Z represent the corresponding times it take to roll down these three
incline planes from top to bottom, then the ratio D_T/T_T^2 is a constant for all, namely
being of the same value for all three inclines as seen in Fig. 20.2. Assuming the same
angle to the horizontal for a set of several different inclines, then smaller-size inclines
should have lower values of distance and time, and bigger-size inclines should have
higher values of distance and time, yet, the ratio D_T/T_T^2 should be identical for all.

Indeed as was shown in the previous chapter, $D_T/T_T^2 = A/2$, and which is constant since acceleration is constant according to Galileo's definition. This can be written more explicitly as:

$$\textbf{DISTANCE}_{\textbf{TOTAL}}/\textbf{TIME}_{\textbf{TOTAL}}^2 = (1/2) * \textbf{Acceleration} = \textbf{constant}$$

It should be noted that the constant $A/2$ here for inclined planes is not $(9.80665)/(2)$ as is the case for freely falling bodies, but rather of much lower value—depending on the angle with the horizontal. The steeper the angle is the stronger the acceleration. The flatter the angle is the weaker the acceleration.

By observing the constancy of the ratio D_T/T_T^2 for all sizes of incline planes, Galileo was able to confirm his hypothesis regarding the proper definition of acceleration. And this was accomplished by eliminating and avoiding speed measurements altogether, as well as by slowing time down to more manageable values that can be measured with 'primitive' water clocks.

Galileo's proportionality strongly resembles Kepler's proportionality regarding his third law as seen in Fig. 12.3, namely the relationship $T^2/D^3 = K$ for all the planets. Interestingly, Galileo's terrestrial analysis of rolling balls, as well as Kepler's formulation of his celestial laws, both are actually referring to similar type of motion involving inertia and gravity, and both were discovered around the same decade, yet, neither Galileo nor Kepler showed any interest in the proportionality discovery of the other, even though they corresponded with each other numerous times! Their correspondence was entirely about specific astronomical issues, with one usually supporting and encouraging the other, as well as about their shared enthusiasm, faith, and devotion to Copernican heliocentricity. They did not correspond about any abstract physical rules and generic laws of motion, because all that was exclusively Galileo's realm; and it was not Kepler's focus in any way. For humanity's scientific progress, their distinct focuses, temperaments, and talents, on the whole, was a fortunate 'division of labor'—and perfectly so! It fell to Newton—one generation later—to fuse the results of both of them into one coherent and systematic physical theory.

Galileo's work could be summarized as follows: before Galileo it was thought that <u>force causes speed</u>, as claimed by Aristotle. Galileo showed that <u>force causes acceleration</u>; that absence of forces results in inertia where speed is maintained steadily; and that the force of gravity which causes all bodies to move downward is a constant force. In other words, a constant force does not lead to constant speed but to constant acceleration.

Chapter 21
Rate of Fall is Independent of Body's Mass

In addition to confirming the above proportionality constant by way of rolling balls down inclined planes, and thus validating his hypothesis regarding the proper definition of acceleration, Galileo also obtained a strong confirmation that the rate of the fall of the ball down an inclined plane does not depend on its mass; observing that heavy balls as well as light balls rolled down the inclines in exactly the same way. Galileo's experiments with pendulums also confirmed this fact, as differences in the weight of the pendulum's bob [with a fixed pair of values for length and angle] did not affect their motions in any way.

This exceedingly important and crucial observation was contrary to what Aristotle had taught, namely that heavy objects fall faster than lighter ones, in direct proportion to weight. Once again, Galileo had to contend and quarrel with ancient and obsolete Aristotelian ideas.

Galileo surmised that if bodies fell in a vacuum, where there was no air resistance to slow some objects more than others, even a feather and a cannon ball would descend at the same rate and reach the ground at the same time. In spite of the fact that differentiated rate of fall is actually often observed for falling bodies due to their distinct weights, Galileo concluded that it was only air resistance that caused lighter objects, say feather or paper, to take longer to fall and reach the ground as compared with heavier objects such as metal balls or rocks, and that if there was no air to interfere, all objects would fall at the same rate. Galileo was able to demonstrate this principle by doing experiments with cannonballs of different weights (where air resistance is only a tiny factor and thus can be ignored), showing that, when dropped from the same height, they reached the ground practically at the same time.

According to a biography by Galileo's disciple and assistant Vincenzo Viviani, Galileo is said to have dropped two spheres of different masses from the Leaning Tower of Pisa, to demonstrate that their time of descent was independent of their mass. Figure 21.1 depicts a drawing of the supposed event. The drawing was reproduced by courtesy of Heritage History at www.heritage-history.com.

While this story has been retold in popular accounts, there is no account by Galileo himself of such an experiment, rather it is believed to be a legend. Most historians

© The Editor(s) (if applicable) and The Author(s), under exclusive license to Springer Nature Switzerland AG 2020
A. E. Kossovsky, *The Birth of Science*, Springer Praxis Books,
https://doi.org/10.1007/978-3-030-51744-1_21

Fig. 21.1 Dropping of
spheres from the leaning
tower of Pisa

believe that it was a thought experiment which did not actually take place, while
Galileo's historian Stillman Drake argues that the event actually did take place just
about as Viviani described it, but that it was a low-key demonstration of what Galileo
knew already for a small group of his students, rather than some high-profile experiment by him with numerous participants where results are not known in advance
and eagerly awaited.

Stillman Drake (1910–1993) was a Canadian historian of science best known for
his extensive work on Galileo. Drake staunchly defended Galileo's experiments as
documented in his published Two New Sciences and in his manuscript notes.

Drake and others worked through Galileo's notes, repeated some of Galileo's
reported experiments, and demonstrated that Galileo was a careful experimentalist
whose observations did play the role in the development of his scientific system
as outlined in Two New Sciences. Drake showed how the complex interaction of
experimental measurement and mathematical analysis led Galileo to his law of falling
bodies. Moreover, Drake deemed Galileo's measurements not only as regulative but
also as constitutive of his science. Other Galilean scholars refrain from going so far,

and claim that Galileo did not perform all the experiments he described in his books and his notes, in part due to the lack of accurate time measurement ability of his era.

In 1961 Thomas Settle at Cornell University in New York succeeded in reproducing Galileo's experiments with inclined planes using the same methods and technologies described in Galileo's writings.

Chapter 22
Galileo's Analysis of Projectile Motion

The general erroneous opinion about projectile motion before Galileo followed largely Aristotelian lines where projectiles were pushed along by an external force which was transmitted through the air. This required the introduction of the concept of the medium, such as air or water, assisting the motion, where parts of the surrounding medium take up the place to the rear of the moving rock or cannonball and push it along. Medieval successors of Aristotle internalized this 'force' in the projectile itself, and coined the term 'impetus' for it. Hence impetus was supposed to cause the object to follow a straight line until it 'lost its impetus', at which point it fell abruptly to the ground along another straight line. Yet, under closer scrutiny, and especially when viewing projectiles traveling through large distances, it becomes obvious that projectiles do not behave in such a manner.

With the unfortunate spread of the use of cannons and artillery in warfare in the 17th Century, the study of projectile motion assumed much greater practical significance, leading to more careful observations. Many observers came to the realization that projectiles did not move in the way Aristotle described; that the path of a projectile did not consist of two consecutive straight lines, but rather that it was instead a singular smooth curve.

Galileo was the first who accurately described projectile motion. He showed that it could be understood by analyzing the horizontal and vertical components of motion separately. No one has gone this analytical path prior to Galileo, and his insight later helped lay the conceptual foundation for Newton's work on the action of forces, and the concept of vectors. Galileo reasoned that what acts vertically on the projectile is the normal tendency of free fall, and that this pulls it towards the earth. But while the object is being pulled down, it is also moving forward horizontally at the same time, and this horizontal motion is uniform and constant according to Galileo's principle of inertia.

© The Editor(s) (if applicable) and The Author(s), under exclusive license to Springer
Nature Switzerland AG 2020
A. E. Kossovsky, *The Birth of Science*, Springer Praxis Books,
https://doi.org/10.1007/978-3-030-51744-1_22

Fig. 22.1 Galileo's drawing
on projectile motion analysis

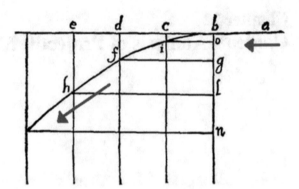

One immediate consequence was Galileo's subsequent discovery that the **parabola** was the theoretical trajectory of a uniformly accelerated projectile in the absence of air resistance. Galileo realized that the parabola is not merely a geometric figure in terms of Conic Sections of the ancient Greeks; but that it has physical manifestations as well in the context of projectile motion. Interestingly, in a very similar way, Kepler borrowed from Conic Sections of the ancient Greeks the **ellipse** for his description of celestial projectile motion of the planets.

In Two New Sciences, Galileo includes his own drawing of projectile motion, as depicted in Fig. 22.1. The insertions of the two arrows in blue color are not of Galileo's; rather they are added by the author in order to aid the reader in following Galileo's line of thought.

Here the horizontal line AB represents a physical flat plane on which an object is carried uniformly from A to B and then continues beyond B over empty space where it starts falling naturally downward due to gravity, tracing the perpendicular BN as well. The object is traversing the points A $\rightarrow$ B $\rightarrow$ I $\rightarrow$ F $\rightarrow$ H, and so forth.

All the segments AB, BC, CD, DE are of equal horizontal length, hence each segment corresponds to an identical duration of time needed to traverse it horizontally, and which shall be arbitrarily defined as the unit of time here. As the body starts falling from point B downwards, the vertical components of the motion for the segments CI, DF, EH, are increasing according to the established relationship $D_T = (A/2) * (T_T)^2$, as discussed in Chaps. 19 and 20. It follows that:

$CI = (A/2)*(1)^2 = (A/2) * \mathbf{1}$

$DF = (A/2)*(2)^2 = (A/2) * \mathbf{4}$

$EH = (A/2)*(3)^2 = (A/2) * \mathbf{9}$

$PQ = (A/2)*(N)^2 = (A/2) * \mathbf{N^2}$ (in general)

In addition to defining parabola in terms of a Conic Section; a parabola can also be described as a graph on the Cartesian Plane of a quadratic function such as Y = X^2, or more generally for X and Y relationship of the form Y = $AX^2 + BX + C$.

Fig. 22.2 Galileo's parabola
deduction for projectile
motion

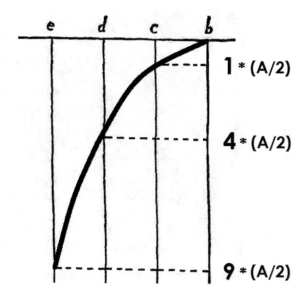

Indeed, the modern projectile motion formula for the vertical height as a function of time T, for a body starting at a certain initial height $H_{INITIAL}$, with zero initial vertical velocity (namely shot directly across with 0° angle horizontally) is:

Height $= H_{INITIAL} - (g/2) * T^2$.

The lower case letter g stands for the gravitational acceleration on the surface of the Earth, which is 9.80665 meter per second squared.

Horizontal inertia implies that horizontal X distance is directly proportional to time elapsed. The motion of the projectile along its horizontal dimension can serve as a perfect clock. In 1 second it would advance 5 meters forward; in 2 seconds it would advance 10 meters forward; in 3 seconds it would advance 15 meters forward; and so forth. Therefore, the above formula for the vertical height can be re-written as $Y = H_{INITIAL} - (g/2)*X^2$, which is exactly the parabola's format.

Figure 22.2 depicts an additional drawing by the author which could perhaps supplement the original one drawn by Galileo in order to better demonstrate the resultant shape of the parabola deduced by Galileo's analysis.

A very recent examination of Galileo's working papers in 1972 revealed the existence of some unpublished documents and personal notes regarding experiments performed by him well before the time of his publication of Two New Sciences. A full account of these documents is given in the Stillman Drake 2000 publication and in the translation of 'Two New Sciences by Galileo Galilei' and appended translation and commentaries titled 'A History of Free Fall—Aristotle to Galileo'. Some of these documents written by Galileo are hard to interpret, and it took over 10 years to decipher much of their contents. These documents generally reveal a great deal of Galileo's previously unknown activities as a mathematician and experimental physicist during his younger years between 1602 and 1609. One of these notes

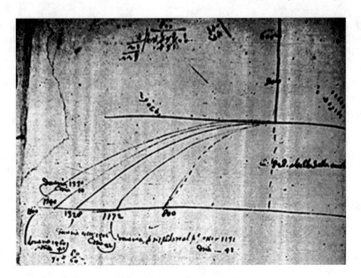

Fig. 22.3 Galileo's drawing of apparent projectile experiment

containing Galileo's informal drawing [Mss. Galileiani, vol. 72, f. 116v] is shown in Fig. 22.3. It appears to relate to actual projectile motion experiments performed by Galileo using an elevated inclined plane situated over a table. This is a kind of investigation that Galileo was not previously known to have undertaken.

In this supposed experiment, Galileo placed an inclined plane on top of a table. The ball thus accelerated via the inclined plane, rolled over the table-top horizontally with uniform motion, falling off the edge of the table as projectile motion, and leaving it with a velocity parallel to the ground, then hitting the floor and leaving a small mark. The mark allowed the horizontal and vertical distances traveled by the ball to be measured. By varying the vertical height of the table, and by varying the ball's horizontal velocity [via changes in the inclined angle or variations in the length of the roll], Galileo might have been able to determine with very high level of confidence that the path of a projectile is parabolic.

Chapter 23
Galileo's Hint at the Concept of the Vector

As Galileo continues to analyze projectile motion in Two New Sciences, he then nearly arrives at the concept of the vector, or rather at the concept of the velocity-vector, stopping just short of defining the force-vector. It is without a doubt that Galileo's Two New Sciences directly paved the way for Newton's Principia half a century later.

Figure 23.1 depicts Galileo's own drawing of superimposed horizontal and vertical constant velocities, resulting in the vector A → C. The insertion of the resultant arrow in blue color is not of Galileo's drawing but rather added by the author for better illustration.

In reference to this drawing, Galileo states his Theorem II, Proposition II:

"When the motion of a body is the resultant of two uniform motions, one horizontal, the other perpendicular, the square of the resultant momentum is equal to the sum of the squares of the two component momenta."

"Let us imagine any body urged by two uniform motions and let ab represent the vertical displacement, while bc represents the displacement which, in the same interval of time, takes place in a horizontal direction. If then the distances ab and bc are traversed, during the same time-interval, with uniform motions, the corresponding momenta will be to each other as the distances ab and bc are to each other; but the body which is urged by these two motions describes the diagonal ac; its momentum is proportional to ac. Also the square of ac is equal to the sum of the squares of ab and bc".

In essence, Galileo calculates the magnitude of vector A → C via the relationship $AC^2 = AB^2 + BC^2$ as per Pythagoras' theorem.

© The Editor(s) (if applicable) and The Author(s), under exclusive license to Springer Nature Switzerland AG 2020
A. E. Kossovsky, *The Birth of Science*, Springer Praxis Books,
https://doi.org/10.1007/978-3-030-51744-1_23

Fig. 23.1 Galileo's hint of
velocity vectors

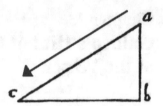

Chapter 24
Galileo's Puzzling Hint at Celestial Application

One truly mystifying short paragraph in Two New Sciences hints possibly at a momentous achievement by Galileo; namely the extrapolation of his terrestrial work on projectile motion to celestial planetary motion. In that captivating paragraph Galileo appears as if suggesting the idea of viewing the planets in terms of projectiles perpetually falling onto and around the Sun. Readers should find the paragraph quite fascinating! The following is the paragraph from Two New Sciences [Crew and deSalvio translation, pp.259–260]:

"SAGREDO: Allow me, please, to interrupt in order that I may point out the beautiful agreement between this thought of the Author and the views of Plato concerning the origin of the various uniform speeds with which the heavenly bodies revolve. The latter chanced upon the idea that a body could not pass from rest to any given speed and maintain it uniformly except by passing through all the degrees of speed intermediate between the given speed and rest. Plato thought that God, after having created the heavenly bodies, assigned them the proper and uniform speeds with which they were forever to revolve; and that He made them start from rest and move over definite distances under a natural and rectilinear acceleration such as governs the motion of terrestrial bodies. He added that once these bodies had gained their proper and permanent speed, their rectilinear motion was converted into a circular one, the only motion capable of maintaining uniformity, a motion in which the body revolves without either receding from or approaching its desired goal. This conception is truly worthy of Plato; and it is to be all the more highly prized since its underlying principles remained hidden until discovered by our Author who removed from them the mask and poetical dress and set forth the idea in correct historical perspective. In view of the fact that astronomical science furnishes us such complete information concerning the size of the planetary orbits, the distances of these bodies from their centers of revolution, and their velocities, I cannot help thinking that our Author (to whom this idea of Plato was not unknown) had some curiosity to discover whether or not a definite "sublimity" might be assigned to each planet, such that, if it were to start from rest at this particular height and to fall with naturally accelerated motion

A. E. Kossovsky, *The Birth of Science*, Springer Praxis Books,
https://doi.org/10.1007/978-3-030-51744-1_24

along a straight line, and were later to change the speed thus acquired into uniform motion, the size of its orbit and its period of revolution would be those actually observed".

"SALVIATI: I think I remember his having told me that he once made the computation and found a satisfactory correspondence with observation. But he did not wish to speak of it, lest in view of the odium which his many new discoveries had already brought upon him, this might be adding fuel to the fire. But if any one desires such information he can obtain it for himself from the theory set forth in the present treatment".

In view of the inquisitional constraint Galileo was under at the time of writing his last book, it is understandable that his treatise on any possible relationship between projectile motion and planetary motion was indirect, guarded, and very limited. Had Galileo discovered any new pattern based on the heretical heliocentric model, he surely could not have explicitly written about it. Galileo does not tell us how information concerning the sizes and velocities of the planetary orbits could be calculated by the reader in order to trace the same computations made by Galileo himself, where he claims to have found a satisfactory correspondence between astronomical observations and his parabolic theory of projectile motion. How could one go about finding any numerical confirmation of Galileo's claim via the application of the data on the six known planets of the era—in conjunction with Galileo's formulas and theories? Galileo's assertion is truly puzzling and vague, but one must remember the highly constrained situation he was under.

Surely Galileo could not go even one tiny step further here other than hinting and suggesting this, and giving exact numerical calculations would surely induce dire inquisitional reaction. Even this modest hint surely was in itself a bold and dangerous undertaking on the part of Galileo, who was of advanced age and in his seventies. But he got away with it. Perhaps Galileo figured that the inquisition book censors in Rome would find reading the mathematical and the geometrical sections of his book quite challenging and difficult in any case, and would not be thorough in checking each and every sentence and paragraph.

Kepler's 1619 publication of Harmonices Mundi containing the third law preceded Galileo's 1638 publication of Two New Sciences by 19 years, and so it is almost a certainty that Galileo was well-aware of Kepler's work at that late stage in his life.

Kepler's third law $T^2 = KD^3$ and its direct consequence

Speed $= [2\pi]/[\sqrt{K}\sqrt{D}]$ are surely distinct from Galileo's

$D_T = (A/2)*(T_T)^2$ expression of purely vertical acceleration, and they are also quite distinct from Galileo's $Y = H_{INITIAL} - (g/2)*X^2$ expression of projectile motion, and so it is doubtful that Galileo is referring directly to Kepler's third law or anything resembling a forth law. Is it possible that Galileo was simply clandestinely attempting to promote and propagate Kepler's third law which was based on the heliocentric model- and which he firmly supported?

Yet, one should keep in mind that planetary motion and terrestrial projectile motion are comparable phenomena. Freely falling bodies and projectile motion experience the same gravitational vector of force on the Earth's surface, and this differs from planetary motions experiencing differentiated gravitational vectors of forces, due to differentiated distances from the Sun. This is the key! Had Galileo truly been able to conceive later in his life of differentiated rates of 'planetary free fall' acceleration, then it might have occurred to him that his parabolic path which always led objects to crush down on the ground and stop, could turn into a circular path, given enough horizontal (inertial) speed and impetus, and which could then be perpetual and smooth.

Newton surely had a firm grasp of Galileo's major contributions and results, and Newton's achievement is based to a large extend on Galileo's work. One wonders whether young Newton had actually read this particular paragraph in Two New Sciences before making his grand discoveries, in which case the elder Galileo can be perceived as if directly guiding and tutoring young Newton, leading him straight towards his final synthesis of all that was discovered before him, and prompting him towards the discovery of classical mechanics.

Chapter 25
Galileo's Discoveries About Pendulum Motion

The pendulum was crucial throughout Galileo's career. Firstly, it served as a tool for timing events, and thus applied as a clock. Secondly, the pendulum served as a fine manifestation of motion to be studied for its own sake, combining the opposing tendencies of inertia and gravitation, and its parameters (such as length, weight, and amplitude) were easily controllable and manipulated, thus enabling the performance of methodical experimental analysis by varying those parameters. Galileo became fascinated with its properties early in his youth. It is said that while attending mass at the cathedral of Pisa—where the seventeen-year-old Galileo carried out his university education in 1582—he noticed the lamps hanging from the ceiling swinging back and forth overhead and became curious. Galileo then compared the time periods of several lamps of approximately the same suspension length, but having different swing spans (i.e. amplitudes) using his heart pulse rate as a 'clock', and he came to the conclusion that they all had just about the same time period. Figure 25.1 depicts the motion of the simple pendulum.

The earliest document about his thoughts on pendulums is a 1602 letter in which Galileo discusses the hypothesis of the pendulum's isochronism. Isochronism is the property of some physical systems to oscillate at a constant frequency regardless of the amplitude of the oscillations. Amplitude is the maximum deviation from the central [rest] configuration, and it is also a measure of the energy embedded in the system. In that letter Galileo claims that pendulums are isochronous, so that the time period is the same for all amplitudes (swing spans), but that so far he has been unable to demonstrate isochronism mechanically.

Galileo was partially right and partially wrong. We now know that simple pendulums are not in general isochronous, although for small amplitude ranges, namely for pendulums with say 0°–40° of displacements from the vertical position, they are indeed nearly almost isochronous. Figure 25.2 is the formula for the period P in time unit of the second; as a function of length L in unit of the meter; and as a function of the amplitude angle θ in unit of the radian [as opposed to the unit of the angle, applying the conversion formula: Radian = Angle*$(\pi/180)$]. The constant acceleration on the surface of the earth is about g = 9.8 m^2. This infinite series diminishes quickly

© The Editor(s) (if applicable) and The Author(s), under exclusive license to Springer Nature Switzerland AG 2020
A. E. Kossovsky, *The Birth of Science*, Springer Praxis Books,
https://doi.org/10.1007/978-3-030-51744-1_25

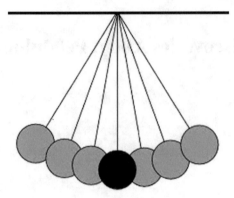

Fig. 25.1 The motion of the simple pendulum

$$P = 2\pi\sqrt{\frac{L}{g}}\left[1 + \frac{1}{16}\theta^2 + \frac{11}{3072}\theta^4 + \frac{173}{737280}\theta^6 + \cdots\right]$$

Fig. 25.2   Period of simple pendulum as a function of L and θ

Fig. 25.3 The
non-isochronism of 3-meter
simple pendulum

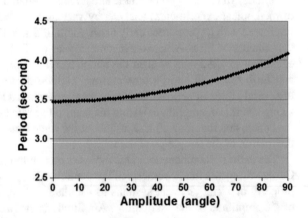

beyond the third term as the fractional coefficients become exceedingly small, hence from the fourth term they could be ignored as nearly zero for all practical purposes.

As an example, for a simple pendulum with length of 3 m, swinging on the surface of the earth, the above formula for the period yields the scatter chart seen in Fig. 25.3. Clearly, for this 3-meter pendulum with amplitudes from 1° to 40°, the period is approximately 3.5 s for the entire variety of amplitudes. For 1° it's **3.476**; for 10° it's **3.483**; for 20° it's **3.503**; for 30° it's **3.537**; and for 40° it's **3.585**. Such minute differentials in time would have been exceeding difficult or rather impossible for Galileo to notice. Only for significantly higher amplitudes the changes in the period are noticeable, such as **3.476** for 1°; **4.088** for 90°; **5.066** for 135°; and **6.833** for 180°.

Exactly the same situation of relative stability in the value of the period exists for any L-length pendulum regarding low amplitudes such as say from 1° to 40°.

The subtle non-isochronism property of the pendulum would be confusing to any experimenter having no prior knowledge of the science, and especially when lacking good timing devices (clocks). Even strong displacements near 90° in the case of Fig. 25.3 would enlarge the period from around 3.5 s to 4.1 s, namely a mere 0.6 s discrepancy. Galileo might have gotten hints about the pendulum's non-isochronism property during some of his experiments, and then perhaps mistakenly blamed the discrepancies on exogenous physical factors, such as latent modes of oscillation, air resistance, or others. The pendulum was perhaps the most challenging and elusive query for Galileo; it had the potential to allow Galileo to break new ground; yet it also had elusive and changeable traits to it that at times threatened to undermine the progress he was making on other fronts. As much of an experimentalist Galileo was, his philosophical approach to nature was very complex and mature, and he was not blindly following observations, or rushing to conclusions. In De Motu Antiquiora manuscript written in his youth, Galileo writes "quaerimus enim effectuum causas, quae ab experientia non traduntur"; and translated from Latin to English it reads: **"experience does not teach us the causes!"** Yet, in the case of the elusive non-isochronism property of the pendulum, such an approach did not serve Galelio well.

Why was the question of the pendulum's isochronism such a crucial issue for Galileo? Simply because the isochronous pendulum, if it truly exists, could then serve as an excellent timing device (clock) for all of Galileo's later work regarding freely falling bodies, balls in inclined planes, projectile motion, and any other types of motion—enabling more precise time measurements. Indeed Galileo used pendulums extensively in his experiments as time measurement throughout his long career. Since all pendulums slow down and eventually stop due to air resistance as well as friction in the string or rope, it follows that in order to be able to apply the pendulum as a clock—it is necessary to have the time period stable while the amplitude is gradually changing and decreasing towards the zero value.

In 1641, the already blind and ailing Galileo conceived and dictated to his son Vincenzo a preliminary design for a pendulum clock which Vincenzo actually began constructing, but had not completed it before his own death in 1649.

The Dutch scientist Christiaan Huygens soon afterwards in 1656 invented the isochronous pendulum clock. Huygens used cycloidal arc restriction on the top of the rope so that the part of the rope that was free to swing was constantly shortened as it swung up to higher angles. A cycloid is the curve traced by a point on the rim of a circular wheel as the wheel rolls along a straight line without slipping. In other words, a cycloid is the elongated arch that traces the path of a fixed point on a circle as the circle rolls along a straight line in two-dimensions. Figure 25.4 depicts the concept of the cycloid.

Fig. 25.4 The cycloid as a
point on a moving circle

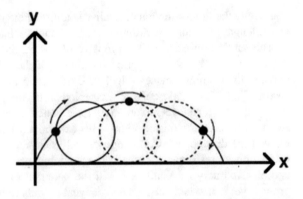

Fig. 25.5' Huygens'
cycloidal pendulum
arrangement

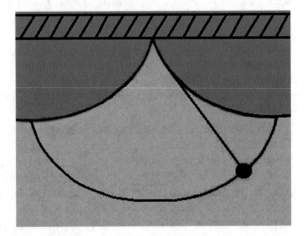

Figure 25.5 depicts the cycloidal pendulum arrangement where the rope originates
exactly between the two cycloid-shaped curves, emanating from the middle top point.

Applying geometrical principles, Huygens deduced that the cycloid curve would
reduce the length for higher angles, and would increase the length for lower angles,
and these two upsetting and cancelling factors would result in having the same period
for all resultant angle-length combinations. In other words, since the period is positively proportional to the length, as well as positively proportional to the angle, as seen
in the formula of Fig. 25.2, it follows that as the cycloidal pendulum gradually slows
down to lower angles, its average length automatically increases, thus compensating
exactly for the reduction in the angle, and consequently the period is perfectly unaf-
fected. It should be noted that the whole arrangement of Huygens' pendulum clock
is based on the fact that a cycloid-constrained pendulum itself follows a cycloid
path, and a somewhat complex mathematical analysis is required to arrive at this
conclusion.

Figure 25.6 depicts a typical 'modern' pendulum-based clock used in many houses
about a generation or two ago.

Fig. 25.6 The
pendulum-based clock

Complicating matters for Galileo was the challenging issue of the 'latent modes of oscillation', namely that the rope itself—if made thick and heavy—constitutes a set of multiple [infinite rather] pendulums of an infinite variety of lengths distributed along the rope. In other words, the rope's segments themselves behave like many mini pendulums distributed along the length of the rope. As an example of an extreme case which illustrates this point, imagine the use of a heavy metal chain made of 8 rings instead of a single light rope. Such an arrangement constitutes 8 different pendulums of distinct lengths tied up together, all competing and struggling to maintain their own unique periods depending on their unique length. The overall curve of the chain of such pendulums continuously changes as it assumes exotic shapes.

In addition, such a complex pendulum with contradictory tendencies quickly loses its energy and stops, well before air resistance and friction get to significantly become factors. The issue of 'latent modes of oscillation' was known to Galileo, and it forced him to use only very thin ropes which for all practical purposes eliminated the problem. By implication, the use of thin ropes limited the weight of the bob Galileo could use, lest the delicate rope were to break as it bears the heavy weight. By implication, the use of light-bob pendulums causes the motions (of such low-energy light pendulums) to slow down rather quickly, because of aerodynamic resistance, and this fact limits the performance of experiments. In Two New Sciences Galileo develops an elaborate theory of the intrinsic tendency of pendulums to slow down and eventually stop due to latent modes of oscillation, regardless of air resistance and friction.

Experiments with multiple pendulums for the purpose of direct comparison with simultaneous swings of pendulums [all moving right next to each other] involve another complication called 'coupling', namely interaction affects, and this phenomenon might have had a negative impact on Galileo's experiments.

One form of coupling is mechanical coupling, which occurs when the horizontal arm of the pendulum structure - from which all the pendulums hang - has some flexibility and therefore it tends to bend due to the forces exerted by the oscillating pendulums.

The other form of coupling is aerodynamic coupling, where the closely-placed pendulums interact via the air media between them, resulting in some mild distortion in their motions.

Pendulums are thoroughly discussed in both Galileo's 'Dialogue Concerning the Two Chief World Systems' and his 'Discourse and Mathematical Demonstrations Concerning Two New Sciences'. Let us summarize the major points Galileo discovered about pendulums in these two books:

A. Pendulums nearly return to their release heights.
B. All pendulums eventually come to rest with the lighter ones coming to rest faster than the heavier ones.
C. The period is independent of the bob weight.
D. The period is independent of the amplitude/angle (partially wrong).
E. The square of the period varies directly with the length.

In the last observation, Galileo is almost correctly discovering the true formula for the period of the pendulum as seen in Fig. 25.2, except for the omission of the angle, but which is approximately correct for low angular values of say those below 40°.

In unit of radian, 40° is converted into $40*(\pi/180)$, namely 0.69813 rad, which is a fraction less than 1, thus, all small degrees less than 40° measured in radian, when squared or powered are even much less than 1 and closer to zero, and in light of the fact that the coefficients 1/16, 11/3072, 173/737,280, and so forth, are of exceedingly low values, therefore the whole bracket containing the angle θ in Fig. 25.2 reduces approximately to 1, and thus can be ignored. The θ bracket is significant only for higher degree values.

For example, for 40°, which is 0.69813 rad, the value of the θ bracket is: $[1+(1/16)*0.69813^2+(11/3072)*0.69813^4+(173/737280)*0.69813^6]=1.03134$, and which is nearly 1.

The formula in Fig. 25.2 is then simplified to the expression $P = 2\pi\sqrt{(L/g)}$ for small angles. Squaring both sides of the equation yields the relationship $P^2 = [4\pi^2/g]*L$, and therefore the square of the period is directly proportional to the length, as Galileo claimed.

Interestingly, it was most likely Galileo who indirectly prompted and inspired Huygens to look into the case of the cycloid in the first place in connection with gravitational fall. Galileo originated the term 'cycloid' giving it its name; popularized the curve; and was the first to make a serious study of the curve.

The study of this curve is a subject rich in mathematical meanings, scientific applications, and important connections within the history of mathematics. While the basic concept of a cycloid is very simple, a variety of more advanced mathematical topics—such as unit circle trigonometry, parametric equations, and integral calculus—are needed for any real mathematical understanding of the topic.

A cycloid is an 'experimental' curve, obtained through a particular 'action' of the rolling of the circle, being one of the first transcendental curves (curves that cannot be derived algebraically). In pure mathematics, two main issues arise immediately with connection to the cycloid. First, it is the length of one cycloidal arc in relation to the size of the circle that generates it, and the second is the area between a cycloidal arc and the straight line its 'generating circle' rolls along. The answer to both these questions puzzled mathematicians for a long time until the development of calculus brought straightforward answers. Indeed, the application of calculus yields surprisingly neat answers: the arc length of a single cycloidal is precisely 8 times the radius of the generating circle, and the area under such an arc is exactly 3 times the area of the generating circle.

According to Galileo's student Evangelista Torricelli, in 1599 Galileo attempted the quadrature of the cycloid (constructing a square with area equal to the area under the cycloid) with an unusual empirical-physical approach that involves tracing both the generating circle and the resulting cycloid on an actual sheet metal, cutting them out and weighing them. Galileo discovered that the ratio was roughly 3:1 but mistakenly deemed the true ratio to be slightly different from the exact 3:1 ratio, and then incorrectly concluded the ratio was an irrational fraction, which would have made quadrature impossible.

Such a novel **empirical-physical approach** for issues in pure mathematics illustrates Galileo's tenacity and persistence in obtaining answers to questions and riddles that troubled him. Nowadays, a lot of what cannot be done via mathematical reasoning is analyzed via computer simulations or methodical computer calculations.

In mathematics and physics, a brachistochrone curve, or curve of fastest descent, is the one lying on the plane between a higher point A and a lower point B, where B is not directly below A, but rather a bit sideways, and on which a bead slides frictionlessly from A down to B under the influence of a uniform gravitational field in the shortest possible time. The brachistochrone problem was one of the earliest problems posed in the calculus of variations. Newton was challenged to solve the problem in 1696, and did so the very next day, concluding that is was the cycloidal arc.

In Two New Sciences, at the end of his treatment of accelerated motion along straight paths, Galileo conjectured that an arc of circumference of the circle is the brachistochrone (a term not used by Galileo). Galileo does not prove his conjecture. The conjecture is false, and we now know that the brachistochrone is the cycloidal arc as Newton demonstrated. Even though Galileo mistakenly arrived at the arc of a circle as the solution instead of the cycloid, yet the entire investigation itself probably had also prompted and inspired Huygens to look into the case of the cycloid in connection with the pendulum.

Paolo Palmieri—in referring to Galileo's unpublished notes [Manuscript 72, at the National Library in Florence, at folio 166 recto]—has conducted a thorough analysis of the manuscript and has plausibly demonstrated that it's most likely an attempt by Galileo to arrive at his brachistochrone conjecture via a reasonable [yet not thorough enough] use of an exhausting and extremely time consuming numerical approach. Palmieri plausibly argues that the manuscript was not about the pendulum's isochronisms, but rather about the brachistochrone problem. The drawing, scripts, and scattered numbers in the unpublished notes indicates that Galileo was calculating the time of descent from A to B via a single straight [sloping] line; then via 2 lines—one which slopes more and one which slopes less than the single line approach—and which yields shorter time interval; via 3 lines which yields even shorter time interval; and finally via 4 lines which is relatively the shortest time interval. But the level of difficulty in such calculations increases dramatically with each new line, so Galileo decided to stop. Galileo concluded decisively that the time of descent shortens with each new added line, and then claimed that the process will converge to a limiting value when the number of lines goes to infinity and the path appears circular.

In spite of terminating with an erroneous conclusion, Galileo's approach to the problem was highly innovative and ingenious.

In a similar fashion as Archimedes who evaluated the volume of the sphere via a method that presaged or foretold of Calculus almost two millennia before Newton and Leibniz—Galileo is foretelling here of calculus as well in a sense. Galileo is dividing and sub-dividing the rectilinear path into smaller and smaller segments, stops from sheer exhaustion, and then proposes the arc of the circle to be the true curve in the limit as the number of linear partitions goes to infinity. In other words, Galileo is viewing the arc as an infinite set of inclined planes. Galileo's persistence in obtaining a solution here leads him to apply an alternative **numerical-experiment approach** to the problem, instead of an actual mathematical analysis which challenged him.

Pierre Duhem (1861–1916) was a French physicist, mathematician, and philosopher of science. According to his radical or unorthodox views he opposed Newton's statement that Principia's law of universal gravitation was indeed deduced from 'phenomena', including Kepler's laws. Somewhat similar criticism had already been given most famously by Immanuel Kant, following David Hume's logical critique of induction. Duhem argued that physics is subject to certain methodological limitations that do not affect other sciences. According to Duhem, an experiment in physics is not simply an observation, but rather an interpretation of observations by means of a theoretical framework. Furthermore (he argues), no matter how well one constructs one's experiment, it is impossible to subject an isolated single hypothesis to an experimental test. Instead, it is a whole interlocking group of hypotheses, background assumptions, and theories that are tested. More generally, Duhem was critical of Newton's description of the method of physics as a straightforward 'deduction' from facts and observations.

Paolo Palmieri has written extensively and eloquently about Galileo's pendulum work and about Palmieri's own extensive attempts at re-producing some of Galileo's pendulum experiments and other results, obtaining equipment, ropes, bobs, and inclined planes, similar to those in Galileo's descriptions. Palmieri's poignant and thought-provoking concluding words are:

> Much has been made, in the philosophical literature, of the relation of empirical facts to theory, and ever since Pierre Duhem's seminal studies, emphasis has been placed on theory-laden phenomena. I find that this emphasis is misplaced. On the scene of experience, I lived-through nothing but the stubbornness of phenomena. The scene of experience is robust. For, try as I might, weird phenomena would resist all sorts of attempts to explain them away. Facts of experience are not easily concocted out of theoretical commitments. You may be struck dumb by the unexpected behavior of artifacts, and wish it to go away by the magic of theory, but you can do nothing about it. You are stuck with it. Galileo's new science erupted out of a willingness to negotiate and trade theoretical norms for stubborn facts of experience, and sometimes, no doubt, the other way round (with consequent discomfort of later scholars). Observations and theory were placed on an equal, unstable footing.

Chapter 26
Galileo's Astronomical Discoveries

Galileo was championing and promoting the heliocentric model throughout his long career. Opposition to the heliocentrism from astronomers and scientists was based in part on doubts due to the absence of an observed stellar parallax with the inefficient technology of that era. Stellar parallax is the expectation of obtaining slightly different vistas of the relatively fixed stars in our galaxy due to the fact that we view them from slightly different position (slightly different angles) within the galaxy during the year as the Earth traces its long orbit around the Sun, moving to extreme 'east' 146,605,913 km from the Sun during Perihelion, and then moving to extreme 'west' 152,589,828 km from the Sun during Aphelion. Since stars appeared the same each season of the year with those antiquated instruments, many astronomers concluded that the Earth stands still, and that it is the Sun that moves.

Heliocentrism was investigated by the Roman Inquisition in 1615 and the institution concluded that heliocentrism was "foolish and absurd in philosophy, and formally heretical since it explicitly contradicts in many places the sense of Holy Scripture".

The German-Dutch spectacle-maker Hans Lippershey invented the description for refracting telescope and filed for a patent in 1608. Galileo improved on Lippershey's description and constructed a telescope with about 3 times more magnification. Later on Galileo constructed improved versions up to about 30 times magnification. The telescope was to prove a great and significant instrument in better understanding the celestial world.

On 7 January 1610, Galileo observed with his telescope what he described at the time as "three fixed stars, totally invisible by their smallness", all close to Jupiter, and lying on a straight line through it. Observations on subsequent nights showed that the positions of these 'stars' relative to Jupiter were changing in a way that would have been inexplicable if they had really been fixed stars. On 10 January, Galileo noted that one of them had disappeared, an observation which he attributed to its being hidden behind Jupiter. Within a few days, he concluded that they were orbiting Jupiter; he had discovered three of Jupiter's four largest moons. Soon afterwards Galileo discovered Jupiter's fourth moon on 13 January.

© The Editor(s) (if applicable) and The Author(s), under exclusive license to Springer Nature Switzerland AG 2020
A. E. Kossovsky, *The Birth of Science*, Springer Praxis Books,
https://doi.org/10.1007/978-3-030-51744-1_26

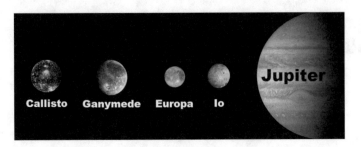

Fig. 26.1 The Galilean Moons—the four largest of Jupiter

The presence of moons in orbit around Jupiter suggested that [even under the Ptolemy's geocentric model] the earth was not the sole center of motion in the cosmos, as Aristotle had proposed, and therefore many astronomers and philosophers initially refused to believe that Galileo could have discovered such a thing. Figure 26.1 depicts the modern description of the four largest moons of Jupiter which are called 'Galilean Moons' in his honor.

On March 13, 1610, Galileo published a short astronomical treatise titled 'Sidereus Nuncius' ('Starry Messenger'). It was the first ever published scientific work based on astronomical observations made through a telescope, and it contains the results of Galileo's early observations of the imperfect and mountainous Moon, the newly observed hundreds of stars that were never before seen with the naked eye, as well as description of his observations of Jupiter's moons. The reactions to Sidereus Nuncius, ranging from appraisal and hostility to disbelief, soon spread throughout Europe. The first astronomer to publicly support Galileo's findings was Kepler, after Galileo explicitly sought Kepler's opinion. Kepler published an open pamphlet in April 1610, titled 'Dissertatio cum Nuncio Sidereo' ('Conversation with the Starry Messenger') where he enthusiastically endorsed Galileo's findings, and offered a range of spec-ulations about the meaning and implications of Galileo's discoveries and telescopic methods, for astronomy and optics as well as cosmology and astrology. But it was not until August 1610 that Kepler was able to publish his own independent confirmation of Galileo's findings, due to the scarcity of sufficiently powerful telescopes. In August that year, Kepler published his own telescopic observations of Jupiter's moons in 'Narratio de Jovis Satellitibus' ('Narration Concerning the Outer Planets' Satellites'), providing further support of Galileo's work. To Kepler's disappointment, however, Galileo never published his reaction to Kepler's 'Astronomia Nova' containing his first and second laws of planetary motion, and which had just been published by Kepler a year before in 1609. Indeed, Galileo quietly or discreetly rejected Kepler's selection of the ellipse as the curve of planetery orbits, because Galileo still believed in shades of Aristotelian aesthetics and the perfection of the symmetrical circle to serve as the shape of the orbits, in spite of being the one who dismantled so much of the false dogmas of the ancient Greeks.

From September 1610, Galileo observed that Venus exhibited a full set of phases similar to that of the Moon. The heliocentric model of the Solar System predicts that

Fig. 26.2   Galileo's Graphical drawings of the phases of Venus

all phases should be visible since the orbit of Venus around the Sun would cause its illuminated hemisphere to face the Earth when it was on the opposite side of the Sun and to face away from the Earth when it was on the Earth-side of the Sun. Making matters a bit easier for Galileo is the fact that Venus orbits the Sun in merely 225 days.

After Galileo's telescopic observations of the phases of Venus, the geocentric model became practically indefensible. Galileo's discovery of the phases of Venus was arguably his most empirical and practical influence contributing to the final transition from the geocentric model to the heliocentric model.

The observation showed that Venus did not remain always between the Earth and the Sun as assumed in the Ptolemaic astronomy and Aristotelian cosmology. Complicating the discovery somewhat was the [false] counter argument that—contrary to the moon which was believed to be dark and merely reflecting the light coming from the Sun—Venus possesses lights of its own, and as it rotates upon its own axis different parts of the planet's surface emanate a diversity of light intensity. The other [false] counter argument was that it is dark indeed, but that its surface is not homogenous and that Venus rotates, thus different parts of it differ in their capacity to retain or reflect light, and this description supposedly explains the resultant differentiation in light intensity.

Figure 26.2 depicts Galileo's own drawings of the observations he made of the phases of Venus, and which was published in his 1623 book 'The Assayer'. The Assayer is also generally considered to be one of the pioneering works of the Scientific Method, bringing up the novel idea that the book of nature is to be read in a mathematical way, rather than as scholastic or as abstract philosophy, as was generally held at the time.

Galileo made naked-eye and telescopic observations of sunspots, and proposed his own theories and analysis of the phenomenon. Their existence raised another difficulty with regards to the dogma of celestial immutability and the perfection of the heavens, as posited in orthodox Aristotelian celestial descriptions, and this led to a controversy. Galileo then wrote a pamphlet in 1612 called 'Istoria e dimostrazioni intorno alle macchie solari e loro accidenti' ('History and Demonstrations Concerning Sunspots and Their Properties', or in short 'Letters on Sunspots'), and it was published in Rome Accademia dei Lincei in 1613. The pamphlet was published as a response to the publication by Christoph Scheiner concerning sunspots and in opposition to Galileo's interpretations, all of which inspired Galileo to write three lengthy letters to counter it. The Accademia wanted to strike a careful balance between introducing extraordinary new ideas and avoiding causing offence to people who might find those views problematic, hence the Accademia tried to persuade Galileo to avoid an aggressive or polemical tone in his letters. The text was presented for censorship to the Roman Inquisition in order to obtain permission to print. The permission was granted only after insisting that Galileo remove from his text his references to the Biblical Scripture and his claims for divine guidance. One of these references that had to be removed was Galileo's claim in the originally submitted manuscript that *'divine goodness'* had led him to advocate the system of Copernicus, and he had to agree to replace it with the phrase *'favorable winds'*. In another reference, Galileo wrote that the idea that the heavens were immutable is 'erroneous and repugnant to the indubitable truth of the Scripture', and like all other mentions of Biblical Scripture, the censors insisted that this too was to be removed.

In the Letters on Sunspots, Galileo claimed that the observations of sunspots were significant in undermining the traditional Aristotelian view that the sun was both unflawed and unmoving. The third letter contains his first clear public statement in support of Copernicus Heliocentrism.

The German Catholic Jesuit friar and astronomer Christoph Scheiner, working independently, also observed sunspots and he announced his discovery at the end of 1611, apparently soon after Galileo's own observations. This led to a long dispute over priority. Scheiner who was wedded to the Aristotelean dogma that the heavenly bodies are unblemished, rejected Galileo's explanation and argued that sunspots are satellites of the Sun—in an effort to save the perfection of the Sun. Galileo correctly argued that the spots are on or near the Sun's surface, and he bolstered his argument with a series of detailed drawings of his observations. Figure 26.3 depicts one of Galileo's own drawings of sunspots in his Letters on Sunspots.

Galileo provides 38 detailed illustrations and geometrical diagrams, which allow the reader to see how his observations relate to his calculations. Galileo argued that if Mercury, the planet closest to the Sun, completes its transit across the backdrop of the Sun in just about six hours; it makes no sense to propose that these sunspots which are much closer to the Sun than Mercury would take around fifteen days to complete theirs, therefore they must be actual part of the surface of the slowly rotating Sun. Indeed we now know that the Sun rotates and spins around its axis once every month approximately. Galileo also argued that the apparent acceleration of the spots as they approach the centre of the Sun and their slower speeds towards the edges, are perfectly consistent with a circular rotation on the surface. The growth in apparent size of the gaps between spots as they approach the centre, and their apparent reduction towards the edges of the Sun, likewise confirm this. Scheiner who was ordained a deacon (slightly lower rank than a priest) and wrote about theology, later wrote his main book 'Rosa Ursina sive Sol' in 1626–1630, about his comprehensive collection of the data

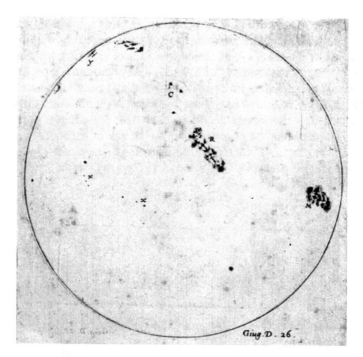

Fig. 26.3   Galileo's Illustration in the Letters on Sunspots

from his observation of sunspots, his analysis, and further mentioning numerous passages and quotations from the Bible and the writings of the church fathers and philosophers to prove that the geocentric view is in accordance with the teachings of the Catholic Church.

In his Letters on Sunspots, Galileo attempts to establish a connection between cosmology and mechanics, and this revolutionary train of thought was most likely the first ever publication of such a daring link in the history of science and astronomy. In addition, this was the first instance in Galileo's works where he (hesitantly) mentioned the concept of inertia, as he attempted to provide an "explanation" of the Sun's motion of rotation. Galileo wrote, "I seem to have observed that physical bodies have physical inclination to some motion."

Galileo observed the Milky Way Galaxy, previously believed to be nebulous, and found it to be a multitude of stars packed so densely that they appeared from Earth to be clouds. Using his telescope he located many other stars too distant to be visible with the naked eye.

Galileo was the first to deduce the cause of the apparent lunar spots and shades as actual lunar mountains and craters. In his study, he also made topographical charts, estimating the height of the mountains. He concluded that the Moon was not what was long thought to have been a translucent, smooth, and perfect sphere, as Aristotle claimed.

Galileo wrote that the tides [on the beaches of the seas and oceans] were caused by the sloshing back and forth of water as the Earth's surface sped up and slowed down because of the Earth's daily rotation on its axis. However, his theory was a failure. If this theory were correct, there would be only one high tide per day. Galileo and his contemporaries were aware of this inadequacy because there are two daily high tides at Venice instead of one, about twelve hours apart. Galileo dismissed the correct idea, held by Kepler, that the Moon caused the tides. Kepler's great insight and continual meditation on planetary motions [as well as his belief in attributing to the Sun a "moving soul" which causes the motion of the planets] led him to the correct view that the tides arise from some sort of 'gravity', as the waters are being pulled towards the Moon.

Chapter 27
Galileo's Trial and Imprisonment

The opposition of the Roman Catholic Church to heliocentrism in general, and to Galileo's version of heliocentrism in particular, sprang from several distinct reasons. Firstly, it contradicted several literal passages in the Bible as mentioned in Chap. 3. Secondly, Galileo's approach to it was accompanied by empirical studies and actual analysis of the sky via his telescopic observations, and which strongly supported the heliocentric model and also demonstrated the newly perceived imperfection and mutability of the heavens. This was considered as the desecration of the supposedly divine sky. Never before were the Moon, the Sun, the planets, and the stars, scrutinized to such a degree via visual analyses; and this created a subtle or subliminal fear among the church establishment that religion itself could be undermined by such activities. Removing the Earth from its central importance as God's focus on people's conduct and placing the Sun at the center of everything also undermined religion in general. Thirdly, the possible ramifications of helicentrism and Galileo's ideas represented a threat to the Catholic Church as the Protestant Reformation began to spread around Europe all throughout the sixteenth century, profoundly affecting Christianity. As wars were breaking out in Europe (supposedly) over the issue of Reformation, the Catholic Church thrived hard to maintain its authority, and clung onto its traditional interpretation of the Holy Scriptures and the geocentric views of Aristotle and Ptolemy much tighter than ever before.

In addition to Galileo who was making groundbreaking celestial observations, Johannes Kepler was also conducting his own independent telescopic observations and research in the field of astronomy, and this only added to the church's anxiety. When Kepler published his book Astronomia Nova in 1609, and forcefully argued in support of heliocentrism, backed by Tycho's meticulous data and his two laws of planetary motion, the church's anxiety became even more intense, and the need to counter these intellectual tendencies became ever more urgent.

© The Editor(s) (if applicable) and The Author(s), under exclusive license to Springer
Nature Switzerland AG 2020
A. E. Kossovsky, *The Birth of Science*, Springer Praxis Books,
https://doi.org/10.1007/978-3-030-51744-1_27

Galileo's publication of Starry Messenger in 1610, together with the included conclusion that it was the Sun at the centre of the universe, had attracted a lot of criticism within the Catholic Church, and consequently Galileo's work had been investigated by the Roman Inquisition in 1615 as possible heresy and perhaps as Protestantism.

In February 1616 Galileo was then ordered by Cardinal Bellarmine's to obey the following verdict:

> ... to abandon completely this opinion, that the sun stands still at the center of the world, and that the earth moves, and henceforth not to teach, hold, or defend this erroneous opinion in any manner whatever, either orally or in writing.

This decree also banned Copernicus' book De Revolutionibus as well as other heliocentric books. Yet, Bellarmine's order did not forbid Galileo from presenting heliocentrism as a purely philosophical and mathematical idea, rather, only its physical interpretation was banned.

Following the church decree in 1616, Galileo avoided any further publication on the topic, hoping that the Church opposition would gradually subside, and he continued his work, never wavering in his firm support for heliocentrism. He was then in his 50s and was suffering from ill health at times.

Suddenly, his long-time friend and supporter Cardinal Barberini, was elected as Pope Urban VIII in 1623. Galileo was delighted, and he then decided that it was the right time to write a very general astronomical book comparing the Copernican and Ptolemaic systems.

Galileo's book, titled "Dialogue Concerning the Two Chief World Systems" was published in 1632, earning formal authorization and approval from the Roman Inquisition and the papal authorities. Pope Urban VIII urged Galileo from the start before he began the writings of the book to give fair and equal arguments for heliocentrism as well as for geocentrism, but not to advocate the heliocentric model exclusively as a physical reality. The new Pope also urged Galileo to express the Pope's own geocentric views on the subject in the book. Galileo fulfilled only the latter request, but not the former request to act as an impartial and neutral judge on the matter.

The Dialogue, as its name suggests, consists of a succession of debates and dialogues between three characters, Sagredo, Salviati, and Simplicio. Sagredo acts as a neutral and unbiased scholar, mediating between Salviati the heliocentrist who argues in favor of Galileo's ideas, and Simplicio the geocentrist who argues against Galileo's ideas.

Whether Galileo deliberately intended it or not, Simplicio often stumbles and gets caught his own errors, and at times comes across as a total fool. Galileo wrote in the preface of the book that the named was respectably and honorably derived from the famous Greek Aristotelian philosopher named (in Latin) Simplicius. Yet indeed Simplicio's derogatory name translates to 'simpleton' in Italian. Hence, the Pope's personal request from Galileo that his geocentrical views be known in the book translated into a portrayal of a silly character called Simplicio representing the Pope in effect. In addition, the Dialogue represented indeed an advocacy book for Copernican heliocentrism and explicit attack on Aristotelian geocentrism.

Galileo was delighted at first, for after years of struggle, hesitation, and caution, his ambition to dismantle in the public domain Aristotelian geocentrism had finally been realized. An entertaining and scientific book in defense of heliocentrism was published for all to see and read, and on the face of it, the book appeared to have earned the formal approval of the Church. Unknown to Galileo, his downfall and suffering were just about to start.

Historians generally agree that Galileo did not act out of malice against his friend the Pope, and indeed Galileo felt quite surprised by the negative reaction to his book on the part of the Church. Galileo was probably so absorbed in his writing, and so enthusiastic about the ability to publish on this essential topic, that he neglected to pay attention to what he promised the Pope, and to be cautious about how he represented him in the book. The Pope was indeed offended for being portrayed and ridiculed as a simpleton and a fool, as well as by Galileo's unfulfilled promise not to advocate the Copernican model as a physical reality. Thus, unintentionally, Galileo had alienated the most influential and powerful supporter in this arena—the Pope himself!

To protect itself, the Church needed to punish and make an example of Galileo, so that others would refrain from taking the same revolutionary path or make new and challenging interpretations of the Bible in other religious and scientific matters. Galileo, who was then nearly 70 years of age and suffering from poor health, was denounced as heretic and summoned to Rome in September 1632. Galileo's difficult and exhausting trip to Rome took him almost five months, and thus his trial began in February 1633.

Galileo was brought before inquisitor Vincenzo Maculani and was formally charged with heresy. Throughout his trial, Galileo maintained that he did not actually advocate heliocentrism, but that he was rather merely discussing it, and he pointed out that the Dialogue had been approved by the Church prior to its publication. However, eventually he was persuaded to admit that contrary to his true intention, any reader of the Dialogue would get the impression that the book indeed was defending heliocentrism as a physical reality.

Galileo's implausible denial that he had ever held Copernican ideas himself after the church ruling on the matter in 1616, and that he never intended to support helio-centrism in the Dialogue, fell on deaf ears. In his final interrogation in July 1633, the judges threatened Galileo with torture if he did not tell the truth, yet he steadfastly denied the heresy charges despite the threat. Figure 27.1 is a dramatic painting which attempts to illustrate the serious character and the menacing atmosphere of Galileo's trial.

The sentence of the Inquisition was delivered on 22 June 1633, and it handed down Galileo the following order: "We pronounce, judge, and declare, that you, the said Galileo… have rendered yourself vehemently suspected by this Holy Office of heresy, that is, of having believed and held the doctrine (which is false and contrary to the Holy and Divine Scriptures) that the Sun is the center of the world, and that it does not move from east to west, and that the Earth does move, and is not the center of the world." He was declared as being guilty of heresy, and he was required to "abjure, curse and detest" those opinions.

Fig. 27.1   Galileo's Trial—Rome 1633

Galileo was sentenced to formal imprisonment at the pleasure of the Inquisition. On the following day, this was commuted to house arrest, under which he remained for the rest of his life. His offending Dialogue was banned, and publication of any of his works was forbidden, including any he might write in the future.

Along with the order came the following penalty: "We order that by a public edict the book of Dialogues of Galileo Galilei be prohibited, and We condemn thee to the prison of this Holy Office during Our will and pleasure; and as a salutary penance We enjoin on thee that for the space of three years thou shalt recite once a week the Seven Penitential Psalms."

According to popular legend, as Galileo walked out of the court room during the last day of his trial after being forced to renounce in front of the court his theory that the Earth moves around the Sun, Galileo allegedly whispered to himself or to his close associates walking alongside him the famous rebellious phrase in Latin "**Eppur Si Muove!**", meaning "**And yet it moves!**" Well, if Galileo's puzzling and perhaps clandestine hint at celestial application of his terrestrial work on projectile motion is understood correctly, as seems in one paragraph of his 1638 publication of Two New Sciences five years later, then it appears that he had indeed the last word on the matter after all!

Chapter 28
Galileo's Writings on Relativity and Infinities

Galileo postulated the main principle of Einstein's Theory of Relativity, namely that the laws of nature are unique and should only be stated once for any frame of reference moving at a constant speed along a straight line. In other words, that the generic laws of physics are independent of the particular speed and direction of the frame of reference of the observing physicist. Admittedly, the motions, speeds, and details of an observed physical system is seen quite differently for any two different observers, say one at rest relative to the system, and another at some motion relative to it, yet, in spite of these differences, both should observe the same laws of motion, such as $F = MA$, and so forth. The implication of this principle is that there exists no absolute motion or absolute rest, and that the concept of motion is relative to the particular observer in question. For example, if two planets are directly approaching each other fast and are on a catastrophic collision course say, then an observer on planet A can rightly claim that his planet is actually at rest while planet B is doing all the moving, while another observer based on planet B can also rightly claim that the situation is totally in reverse, and that his planet B is actually at rest and that the other planet A is doing all the moving. There exists no physical experiment that can be performed to prove or disprove any one of these two disagreeing observers on planets A and B. This essential principle in physics is indeed the framework which Einstein's Special Theory of Relativity is based on.

Galileo was the first to use thought experiments, three centuries before Albert Einstein became well-known for applying them. An instructive and very interesting example of such thought experiment given by Galileo relates to the acceleration of falling bodies, which Galileo stated that it must be independent of the weight or the mass of the bodies. Let us suppose that Aristotle's theory is indeed correct (Galileo argued), so that heavy bodies fall much faster than light bodies. Let us imagine then two blocks of equal weights, about to be dropped next to each other, in such a way that the blocks are connected horizontally via a slender and very weak thread or string. Under such scenario, when we drop the blocks together while officially declaring them as one singular entity, the expectation is that they should fall rapidly as a single heavy big block. Yet, if we perform the experiment a bit differently and simply cut the thread before dropping them, while officially declaring them as two distinct blocks,

A. E. Kossovsky, *The Birth of Science*, Springer Praxis Books, https://doi.org/10.1007/978-3-030-51744-1_28

then according to Aristotle's theory the two blocks should fall more slowly (being of half the weight of the combined arrangement, and thus each consisting of lighter mass), and which is an absurd conclusion. How could the weak and slender thread or our own verbal declarations regarding whether they fall together or fall individually make any difference in the speed of the fall?! It follows that Aristotle's idea must be a fallacy, and that bodies fall at the same rate and at the same speed regardless of weight.

Galileo's other thought experiment foretold of Einstein's ideas of Special Relativity, and it established what is known as Galilean Relativity. Opponents of heliocentrism argued against the possibility that the Earth rotates or that it orbits around the Sun, based on the (misguided) assumption that things that are not nailed down or fixed in place such as the clouds in the sky or flying birds should be drifting and be left behind, namely that they should not be joining in the Earth's motion. Their argument was that since clouds and birds and all other loose things actually do move within the normal framework of the Earth as we all observe daily, it must be then that the Earth is standing still. Galileo's thought experiment imagines the hold of a large ship that moves smoothly and directly in a straight line and with uniform and steady speed across the ocean. The 'hold' of a ship is an enclosed windowless space within the ship designed typically for storing cargo, and passengers who happened to venture inside the hold cannot really tell whether the ship is stationary or moving simply by 'looking' outside at the water to compare, since there are no windows. When passengers in the hold perform various physical experiments in an attempt to verify or disprove the ship's movement, they always fail. As an example, passengers may throw small rocks straight up vertically, and perhaps expecting them to trail behind and lag if the ship is really moving forward, yet this never happens even when the ship is in motion, and these rocks always fall straight down in the same place so that the passengers could easily catch them with their hands at the same spot. They could also observe water dripping straight down, not lagging or trailing behind. Aristotelian physics predicts that the rocks should drift back and lag behind if the ship is moving forward, but experience shows that this is never the case, and that all physical experiments in the hold of the ship results in the same way whether the ship is moving relative to the water or stationary. In conclusion, there exists no experiment that could 'determine' whether the ship is in motion or at rest, and this is so because motion itself is relative, not absolute.

In addition, Galileo wrote about the possibility of differentiated levels of infinities - centuries before the German mathematician Georg Cantor (1845–1918) singlehandedly developed the subject regarding the Hierarchy of Infinities. Galileo wrote about a particular paradox in order to demonstrate one surprising fact about infinite sets. In Two New Sciences, Galileo points out to the fact that while some integers are squares (such as $2 * 2 = 4$, $3 * 3 = 9$, $5 * 5 = 25$); other integers are not (such as 7, 11, 14); and thus he concludes that the set of all the integers joined together, namely both the squares and the non-squares, must constitutes a more numerous set than the set of the squares (and also more numerous set than the set of the non-squares). Hence

Galileo is able to demonstrate that two distinct infinite sets, namely, the entire set of all the integers, and the 'smaller' set of all the squares, can be considered (perhaps) as being of unequal sizes!

But then Galileo argues to the contrary, namely that these two sets should be considered as of equal size, simply because of the fact that for every square there exists exactly one integer (its square root), and because of the fact that for every integer there exists exactly one square (it's square); therefore each element in the first set points to a unique element in the other set, so that they must be of equal sizes. Galileo's argument here applies the concept of the 'one-to-one correspondence' in the context of infinite sets, a concept which would be formalized later by Cantor.

These questions and dilemmas were discussed briefly in a less formal manner, and with less detailed arguments, well before Galileo, yet Galileo's name became associated with these issues after his publication of Two New Sciences.

Galileo's conclusion was that concepts such as equal, less, and greater, can only be applied to finite sets and not to infinite sets. In summary, Galileo warned that one cannot talk of infinite sets and claim that one infinite set is truly greater or less than another infinite set, and that there exists only one unique 'size' for infinite sets, and that size is called 'infinite'.

In the nineteenth century the mathematician Georg Cantor created set theory, which has become a fundamental theory in mathematics. Cantor's work demonstrated that it is possible to define valid comparisons between infinite sets in a meaningful way, and that as a consequence some infinite sets are indeed larger than others. According to Cantor's definition of sizes of the infinite, the two sets that Galileo considered, namely the set of all the integers and the set of all the squares, are indeed of the same infinite 'size', while the infinite set of all the real numbers, including fractional and irrational numbers, such as all the points on a 1-meter ruler for measuring distance say, are considered by Cantor as 'larger' than the 'smaller' infinite set of all the integers.

Chapter 29
Galileo's Work on Sound and Speed of Light

Galileo is lesser known for, yet still credited with, being one of the first to understand sound frequency. By scraping a chisel at different speed, and by scratching the metal part of knife blade at different spacing scheme, Galileo linked the pitch of the sound produced to the spacing of the chisel's skips, a measure of frequency.

In 1638, Galileo described an experimental method to measure the speed of light by arranging that two observers, each having a lantern (lamp) equipped with shutters, observe each other's lanterns at some distance. The first observer opens the shutter of his lamp, and, the second, upon seeing the light, immediately opens the shutter of his own lantern. The time between the first observer's opening his shutter and seeing the light from the second observer's lamp indicates the time it takes light to travel back and forth between the two observers. Galileo reported that when he tried this at a distance of less than a mile, he was unable to determine whether or not the light appeared instantaneously.

Today we know that the reason why Galileo was not able to observe the speed of light over such a short distance is that light travels way too fast for such an experiment. Light travels at 186,282 miles per second; hence it takes only 0.00001073 of a second to traverse these two miles.

© The Editor(s) (if applicable) and The Author(s), under exclusive license to Springer Nature Switzerland AG 2020
A. E. Kossovsky, *The Birth of Science*, Springer Praxis Books,
https://doi.org/10.1007/978-3-030-51744-1_29

Chapter 30
René Descartes—The Rationalist Mathematician

René Descartes (1596–1650) was a French philosopher, mathematician, and scientist. Descartes is widely regarded as one of the founders of modern western philosophy. Much of subsequent western philosophy is a response to his writings which are still studied closely to this day. Descartes' 'Meditations on First Philosophy' continues to be a standard text at most university philosophy departments. Descartes was also one of the key figures in the Scientific Revolution, frequently setting his views apart from those of his predecessors. Figure 30.1 is a portrait of Descartes—Source: The Free Library.

Descartes' influence in mathematics is even more evident and profound, and the Cartesian coordinate system as depicted in Fig. 30.2 is named after him. He is credited as being the father of analytical geometry, which is the bridge between algebra and geometry, and which was used in the discovery of infinitesimal calculus and analysis. Descartes' analytical geometry is in essence 'a way of visualizing algebraic formulas'. Besides the essential applications of the Cartesian plane in pure mathematics, people nowadays have very little awareness or appreciation of how much Descartes had impacted their daily lives with this invention! We take graphs and charts for granted, as if they were always there for us to use from time immemorial, as if the whole concept is so very obvious, yet it was Descartes who endowed us with it. It is very hard for us nowadays to imagine life before the seventeenth century where nobody whatsoever in the whole world was drawing or reading any chart or graph! When economists draw demand and supply curves on the price versus quantity chart to arrive at the equilibrium market price; when cardiologists chart heart failure incidents versus cholesterol levels for a large population; when engineers analyze the strength of material in a stress versus strain chart; they all owe gratitude to Descartes.

It should be noted however, that Descartes developed his coordinate system only to demonstrate the link between geometry and algebra, and that it is most probable or rather a certainty, that he never imagined it being used in later generations to plot the relationship between two real-world variables, namely as a statistical tool!

When Descartes was one year old, his mother Jeanne Brochard died after trying to give birth to another child who also died. René lived with his grandmother and

© The Editor(s) (if applicable) and The Author(s), under exclusive license to Springer Nature Switzerland AG 2020
A. E. Kossovsky, *The Birth of Science*, Springer Praxis Books,
https://doi.org/10.1007/978-3-030-51744-1_30

Fig. 30.1 René Descartes

Fig. 30.2 The Cartesian
coordinate system

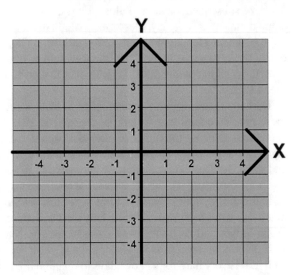

with his great-uncle. At college he was introduced to mathematics and physics, including Galileo's work. Descartes studied Civil Law at the University of Poitiers, in accordance with his father's wishes that he should become a lawyer.

In his book 'Discourse on the Method', Descartes recalls:

> I entirely abandoned the study of letters [Civil Law]. Resolving to seek no knowledge other than that of which could be found in myself or else in the great book of the world, I spent the rest of my youth traveling, visiting courts and armies, mixing with people of diverse temperaments and ranks, gathering various experiences, testing myself in the situations which fortune offered me, and at all times reflecting upon whatever came my way so as to derive some profit from it.

Given his ambition to become a professional military officer, Descartes joined as a mercenary the Protestant Dutch States Army in Breda in 1618 under the command of Maurice of Nassau. Later Descartes was in the service of the Catholic Duke Maximilian of Bavaria since 1619, and was present at the Battle of the White Mountain outside Prague in November 1620. One finds it very hard to reconcile within the personality of this great and exceptional man two diametrically opposing traits: the malevolent tendency towards the Martial Arts and senseless human slaughter on the battlefield, together with the benevolent and noble tendency towards exceptional honesty and totally open-minded intellectual inquiry. It was after a series of dreams or visions, and after meeting the Dutch philosopher and scientist Isaac Beeckman who sparked Descartes' interest in mathematics and physics, that Descartes concluded that his true path in life was the pursuit of wisdom and science.

According to Adrien Baillet, on the night of 10–11 November 1619, while stationed in Neuburg an der Donau, Germany, Descartes shut himself in a room with a masonry heater to escape the cold. While resting there he had three dreams and believed that a divine spirit revealed to him a new philosophy. Upon exiting, he had formulated analytical geometry and the idea of applying the mathematical method to philosophy. He concluded from these visions that the pursuit of science would prove to be for him the pursuit of true wisdom and a central part of his life's work. It was those experiences and occasions that led Descartes to state his famous slogan "**I think, therefore I am**".

Descartes then worked on free fall, Conic Sections, and fluid statics, and believed that it was necessary to create a method that thoroughly linked mathematics and physics. In 1620 Descartes left the army. In 1627 Descartes met with several other scholars and alchemists, and was urged by Cardinal Bérulle to write an exposition of his own new philosophy in some location beyond the reach of the Inquisition. Descartes returned to the Dutch Republic in 1628, joined the University of Franeker, and later also joined Leiden University studying mathematics.

In Amsterdam, Descartes had a relationship with a servant girl, Helena Jans van der Strom, with whom he had a daughter, Francine, who was born in 1635. His daughter subsequently died in 1640 due to scarlet fever, at the very young age of five. Russell Shorto postulated that the experience of fatherhood and the heartbreak of losing a beloved child formed a turning point in Descartes' work, changing its focus to a quest for some universal answers to questions such as the meaning of life, suffering, and existence.

Despite frequent moves, Descartes wrote all his major work during his 20 plus years in the Netherlands. In 1633 Galileo was condemned by the Catholic Church, and this prompted Descartes to abandon plans to publish 'Treatise on the World', his work of the previous four years. Nevertheless, in 1637 he published part of this work in three essays: Les Météores (The Meteors), La Dioptrique (Dioptrics), and La Géométrie (Geometry), preceded by an introduction—his famous essay Discours de la méthode (Discourse on the Method).

In Discourse on the Method, Descartes rejected the notion that everything could be determined by pure logical analysis, without recourse to observation or experiment. Descartes lays out four rules of thought, meant to ensure that our knowledge rests upon a firm foundation:

> Instead of the great number of precepts of which Logic is composed, I believed that I should find the **four** which I shall state quite sufficient, provided that I adhered to a firm and constant resolve never on any single occasion to fail in their observance.

1. *Doubt everything.*

> The first of these was to accept nothing as true which I did not clearly recognize to be so: that is to say, carefully to avoid haste and prejudice in judgments, and to accept in them nothing more than what was presented to my mind so clearly and distinctly that I could have no occasion to doubt it.

2. *Break every problem into smaller parts.*

> "The second was to divide up each of the difficulties which I examined into as many parts as possible, and as seemed requisite in order that it might be resolved in the best manner possible."

3. *Solve the simplest problems first.*

> The third was to carry on my reflections in due order, commencing with objects that were the most simple and easy to understand, in order to rise little by little, or by degrees, to knowledge of the most complex, assuming an order, even if a fictitious one, among those which do not follow a natural sequence relatively to one another.

4. *Be thorough.*

> The last was in all cases to make enumerations so complete and reviews so general that I should be certain of having omitted nothing.

Descartes continued to publish his work concerning both mathematics and philosophy for the rest of his life. In 1641 he published a metaphysics work, Meditationes de Prima Philosophia (Meditations on First Philosophy), written in Latin and thus addressed to the learned. It was followed, in 1644, by Principles of Philosophy, a kind of synthesis of the Discourse on the Method and Meditations on First Philosophy. In 1643, Cartesian philosophy was condemned at the University of Utrecht as heresy, and Descartes was obliged to flee to the city of Hague for his safety.

By 1649, Descartes had become famous throughout Europe for being one of the continent's greatest philosophers and scientists. That year, Queen Christina of Sweden invited Descartes to her court in order to organize a new scientific academy and to tutor her about his general philosophical ideas. Descartes accepted, and moved to Sweden in the middle of the harsh winter. He was a Catholic coming to a Protestant nation. Soon it became clear that they did not like each other, and they mostly stopped their interactions and studies. On 1 February 1649 Descartes contracted pneumonia and passed away on 11 February 1649.

Descartes is regarded as a thinker who emphasized the use of reason to develop the natural sciences. Descartes established the possibility of acquiring knowledge about the world based on deduction and perception. He was aware that experimentation was necessary to verify and validate theories. Yet Descartes distrusted sensory evidence as much as he avoided any undisciplined flirtations of abstract thought.

Descartes also rejected any appeal to final ends or purposes, divine or natural, in explaining natural phenomena, and instead emphasized the mechanical and causality aspects of nature.

His attempt to ground theological beliefs on reason, instead of blind faith, encountered intense opposition in his time. Descartes arguably shifted the authoritative guarantor of truth from God to humanity. With Descartes, the human being is now given a measure of self-responsibility; each person is turned into a reasoning adult, an emancipated being equipped with autonomous reason, as opposed to a child obedient to God.

This was a revolutionary step that established the basis of modernity, the repercussions of which is still being felt nowadays.

In 1663, fourteen years after his death, the Pope in Rome placed all Descartes works on the Index of Prohibited Books.

Descartes laid the foundation for seventeenth century continental rationalism, later advocated by Baruch Spinoza and Gottfried Leibniz, and opposed by the empiricist school of thought led by John Locke and David Hume.

Led by Descartes, philosophers had begun to formulate a new conception of nature as complex, impersonal, and lifeless machine. Yet for some in the universities of Europe, all this might well have never happened, as they continued to hold onto outmoded Aristotelianism, which rested on the geocentric view, emphasizing humanity's supposed centralism and importance in the universe, and dealt with nature in qualitative rather than quantitative terms.

Descartes was the first mathematician to use the notation where the lowercase letters at the beginning of the alphabet (such as a, b, c) represent *known* data or fixed values, and the letters at the end of the alphabet (such as x, y, z) represent variables or *unknowns*. This has been adopted as the modern standard. Descartes' significant contribution in this context was his very modern treatment of independent variables. Descartes' understanding of algebra was profound, and he contributed to the understanding of polynomials and their roots. One example of a single indeterminate polynomial is $Y = 5X^7 + 2X^6 + 4X^3 + 9X + 12$. The root of a polynomial is some possible value of X which yields 0 for Y. Descartes discovered that the number of distinct roots of a polynomial is equal to the degree of the polynomial (the value of the highest exponent or power, which is 7 in the above example). Descartes was willing to consider negative roots which he called 'false roots', as well as imaginary roots, involving the square root of -1.

Descartes also pioneered the standard notation and convention that uses superscripts to express powers or exponents.

For example, the use of the number 2 in the expression X^2 indicating squaring [X to the power 2] was adopted by Descartes.

Descartes revolutionized the study of mathematics by joining its two major fields—algebra and geometry. The Cartesian coordinate system was developed to locate points on a plane but it evolved into what we call analytic geometry. The

Fig. 30.3 Circle, Ellipse, Parabola, and Line in Cartesian plane

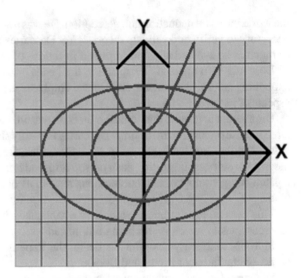

fundamental principle of analytic geometry can be described as 'all pairs of values satisfying the equation are coordinates of points on a curve; and, conversely, all points on the curve have coordinates which satisfy the given equation'.

Figure 30.3 visually and geometrically demonstrates the relationships between variable **Y** and variable **X** via the Cartesian Plane for the following four algebraic equations:

$$X^2 + Y^2 = 4 \qquad \text{for the circle}$$
$$X^2/16 + Y^2/9 = 1 \text{ for the ellipse}$$
$$Y = 2X - 2 \qquad \text{for the line}$$
$$Y = X^2 + 1 \qquad \text{for the parabola}$$

For example, the circle is the set of all joint values (points) of X and Y that satisfy the equation $X^2 + Y^2 = 4$. The point $X = 2$ and $Y = 0$ satisfies the equation. The point $X = 0$ and $Y = 2$ satisfies the equation. The point $X = -2$ and $Y = 0$ satisfies the equation. The point $X = 0$ and $Y = -2$ satisfies the equation. The point $X = \sqrt{3}$ and $Y = 1$ satisfies the equation. And so forth. Taken together, the set of all these points form a circle on the Cartesian Plane.

Probability density functions and histograms in statistics are based wholly on the Cartesian coordinates system, and so do numerous other graphical representations of data in Descriptive Statistics.

Descartes' work provided the basis for the Calculus developed by Newton and Leibniz, who applied infinitesimal analysis to the tangent line problem (i.e. derivative $\Delta y/\Delta x$), thus permitting the evolution of that branch of modern mathematics, and facilitating a big part of Newton's work in physics as well. Figure 30.4 which depicts two distinct tangents line shown at $(-2, -3)$ and at $(3, 3)$, reminds one of the inquiry

Fig. 30.4 The tangent or
slope of a curve in Cartesian
plane

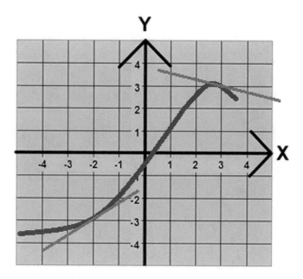

that naturally and immediately arises regarding the tangent/slope/derivative at any
point for any given geometrical curve, and which algebraically translates into $\Delta y/\Delta x$.

In that sense regarding the Cartesian Plane, Descartes had tremendous and direct
influence on young Newton, and this is arguably one of Descartes' most important
contributions to the birth of science.

Chapter 31
Isaac Newton and the Birth of Modern Physics

Isaac Newton (1642–1727) is the last key figure in this story, and the person who completed the Scientific Revolution, bringing to fruition all that was accomplished before him in the previous generations since Copernicus. It was Newton who unified the terrestrial work of Galileo and the celestial analysis of Kepler into one coherent and consistent scientific theory regarding the laws of motion that has stood the test of time. Figure 31.1 depicts a portrait of Isaac Newton.

Newton was an Englishman who contributed greatly to mathematics, physics, and astronomy, yet, he also wrote clandestinely semi-heretical pamphlets about theology, and experimented in alchemy. Newton is widely recognized as one of the most influential scientists in history. His towering and colossal book 'Principia' laid the foundations of classical mechanics, and together with his three laws of motion and universal gravitation, dominated the view of the physical universe for generations to come without modification, until Einstein came along. Newton also contributed significantly to optics, and he shares credit and priority with Gottfried Wilhelm Leibniz of Germany in developing the calculus. Newton realized early in his research that calculus constitutes a body of mathematical knowledge essential for properly stating the laws of physics and for deriving the numerous applications of these laws, and that realization gave him the motivation to develop calculus in the first place.

By being able to derive Kepler's three laws of planetary motion directly from his mathematical description of gravity and motion, and by using the same laws to explain the tides, the precession of the equinoxes, the trajectories of comets, and several other celestial phenomena, Newton then removed any doubts about the validity of Copernicus' heliocentric model of the Solar System. Moreover, Newton was able to demonstrate that terrestrial motion on Earth (such as free falling bodies and projectile motion) as well as celestial motion (such as planets, comets, and the Moon) could all be explained and be accounted for by the same set of physical laws that he formulated.

Newton managed to build the first reflecting telescope. In addition, he developed a complex theory of color after personal observations and experiments with prisms, decomposing the (apparently) white color coming from the Sun into the spectrum of

© The Editor(s) (if applicable) and The Author(s), under exclusive license to Springer Nature Switzerland AG 2020
A. E. Kossovsky, *The Birth of Science*, Springer Praxis Books,
https://doi.org/10.1007/978-3-030-51744-1_31

Fig. 31.1 Isaac Newton

the visible colors. The entirety of Newton's extensive work on light and colors was published much later in 1704 as a book titled 'Opticks'. The book became highly influential in the theoretical study of optics, as well as for its practical applications. In addition, Newton figured out the theoretical speed of sound based on calculations of air compression and rarefaction. Even though his overall theoretical approach was correct, yet his derived speed had 20% discrepancy with empirical measurements of the speed of sound. That discrepancy was corrected and adjusted more than a generation later by French physicists who modified his approach slightly to arrive almost exactly at the speed of sound. In mathematics, besides his contribution to calculus, Newton also studied power series, invented a method for approximating the roots of a polynomial, enlarged the binomial theorem to non-integer exponents, and contributed to the study of cubic plane curves, among other contributions.

Newton had spent most of his life immersing himself with intense intellectual activities, and that effort was only partially related to mathematics and physics, while a good portion of it was spent on theological and biblical studies, as well as alchemical research into all sorts of unknown or little-understood substances which Newton was extremely curious about. Alchemy is the medieval forerunner of chemistry. This alchemy research was almost always accompanied with the customary (and risky) tasting and smelling of the chemicals, as was the norm at that era.

Isaac Newton was born on 25 December 1642 (i.e. on Christmas Day according to the Julian calendar which was in use in England at the era). The place of birth was Woolsthorpe in England. Newton's father, who was also named Isaac Newton as his son, was a modest farmer of a small estate at Woolsthorpe, and unfortunately for his future son he died about three months before Isaac was born. Isaac, who was born prematurely and without a father, was quite small and weak as an infant, and thus he was not expected to survive. Yet he did, in spite of the odds, and he lived to the ripe old age of 84. Unfortunately for the young boy, his mother remarried another man while Isaac was just three years old, and then she went to live with her new husband. She left little Isaac behind at Woolsthorpe and entrusted him with the care of his maternal grandmother, while his mother bore three more children in her second marriage. Young Isaac disliked his stepfather a great deal, either for just being a stepfather as opposed to a biological father, or perhaps due to the stepfather's treatment and behavior toward the young boy. Young Isaac also felt some antagonism or profound disappointment with his mother for remarrying and for leaving him behind, as it induced within him some feelings of being abandoned. Yet, the mother did not actually abandon Isaac. In any case, from the time of her new marriage to the death of his stepfather, which were nine years in total, little Isaac was separated from his mother for the most part, from the age of three to the age of twelve. Newton's occasional negative emotional tendencies in the social context later in his life, his excessive suspicion about the motives of his associates, colleagues, and friends, are easily explained and ascribed to that traumatic period in his childhood when he was separated from his mother, and the way he reacted to it as a child.

That sense of insecurity and suspicion acquired in his childhood, caused him later in life to feel obsessively anxious and untrusting when his work had to be published, and especially when it was necessary to defend his ideas when the relevant scientists or mathematicians criticized the content of his writings. Such were the negative thoughts and emotions that accompanied Newton throughout his life, and which can be easily traced to the trauma of his childhood.

Newton entered The King's School Grantham, where he was educated from the age of twelve until he was seventeen. The school taught Latin and Greek, as well as provided for a solid foundation in general mathematics. Unfortunately again for young Isaac, his mother became a widow for a second time after his stepfather passed away, and his mother returned to Woolsthorpe. In October 1659 she removed Issac from the school during his last year there, so that he could be with her at Woolsthorpe, and she attempted to make young Isaac a farmer at their modest estate. Fortunately for science, young Newton hated farming! He could not bring himself to concentrate on the harvest, fruit trees, cattle, and the affairs of the estate, finding himself instead resting absentmindedly under apple trees, reading books and contemplating issues in science. Finally, Henry Stokes, the master at King's School, who was already highly impressed with Isaac's scholastic talents and his strong enthusiasm to study further, was able to persuade Isaac's mother to heed her son's desires and likes, and to send him back to school. Issac was then able to complete his education there.

Newton entered Trinity College Cambridge in 1661, but had to pay for his studies by performing some duties there. In 1664 he was awarded a scholarship which guaranteed him the financial basis he needed for the next four years until his graduation. By now, Newton was developing mathematical and scientific ideas well beyond his standard curriculum at Cambridge, further reading the work of modern astronomers such as Kepler and Galileo. Particularly, the philosophy of Descartes had profound influence on young Newton, as he read the 1656 Latin edition of Descartes's volume Opera Philosophica, which contained also the Meditations, the Dioptrics, and the Discourse on Method. Yet, Newton had confined to himself all the progress of his studies and all his new ideas, scribing them in his own notebooks, without sharing them with others.

In August 1665 Newton had obtained his bachelor degree at Cambridge, and then suddenly the university closed as a precaution against the Great Plague which struck England again. Newton returned home to Woolsthorpe for the next two years, contemplating at leisure what he had learned before, synthesizing and analyzing all the knowledge that he had gained. It was during these plague years that Newton developed his ideas about light and colors and wrote an essay titled 'Of Colours'. This essay later led to his larger and more detailed publication of 'Opticks'. It can be said that by the end of 1666, young Newton had become the de facto leading mathematician in Europe and by extension in the whole world, with his knowledge of calculus and several other essential and innovative mathematical topics.

It was while staying at Woolsthorpe in the last months of 1666 that Newton had his great insight about forces being defined as the mass multiplied by the acceleration, and about the inverse square relation for gravitational force, so that the force acting on a planet decreases in proportion to the square of its distance from the Sun, and how all this fit nicely with his analysis of the Moon's motion, with Kepler's laws for the planets, and with Galileo's experiments. Newton, now only 25 years old, was able to accomplish these great feats at such a young age.

As the Plague subsided, Newton returned to Cambridge after it re-opened in 1667, and was subsequently elected to a fellowship in Trinity College. In 1669 he became a professor of mathematics at the college, and he elected the topic of optics and his own research about it as his initial topic for his lectures. Around that time Newton constructed the first ever reflecting telescope, and began writing a long treatise of his account on calculus which was completed around 1671, but not published. His mathematical lectures later incorporated also the topic of algebra.

Interestingly, each of the three main protagonists in the birth of science, namely Kepler, Galileo, and Newton, constructed their own telescopes and contributed greatly to the study of optics, while the other related protagonist in this history, namely Rene Descartes, also contributed significantly to optics.

Initially, in the sixteenth century and early in the seventeenth century, the focus of the physical science was of motion itself, free fall, inertia, the motions of the Moon and the planets. Soon afterwards, the second focus and another vital activity of the Scientific Revolution was optics, or rather the nature of light itself. It started with Kepler's publication of his major work on optics in 1604, and continued with Descartes' additional discoveries in optics and his statement of the sine law of refraction, relating the angles of incidence and the emergence at interfaces of the media through which light passes. Descartes' discovery added a new mathematical regularity (pattern) to the study of light, and this discovery together with the mathematical regularities discovered by Kepler, Galileo and Newton, supported the general notion that the universe is constructed according to exact mathematical patterns, and that it is very much mechanical and predictable, as opposed to the fuzzy, qualitative, and philosophical notions of the ancient Greeks. But if motion was challenging enough or a tough puzzle to solve for the minds of the great thinkers of the late Renaissance period, then light was even more of a mystery and a challenge. Light moves, and it moves fast, and one cannot ask it to rest for a moment so that its nature can be examined and analyzed by the scientists. As shall be discovered by later generations of

Fig. 31.2 Newton dispersing sunlight into colors

physicists, light is a thread that runs through thermodynamics, through electromagnetism, and it is one of the main themes in Einstein's Theory of Relativity. Descartes placed light as central to his mechanical philosophy of nature, arguing that light is simply motion transmitted through a material medium (air or water) as a <u>wave</u>. Newton accepted the idea of the mechanical nature of light wholeheartedly, yet he believed in the atomistic explanation of light, and he argued that light consists of little material <u>corpuscles</u> that are in motion.

Newton's corpuscular theory of light was speculative, but his core contribution to light was about colors. Newton performed a series of experiments around 1666 in which the spectrum of a narrow (white) sunlight beam was projected onto a wall of a darkened chamber, and then dispersed through a prism into its multi-colored spectrum. Figure 31.2 depicts young Newton performing the experiment of the dispersion of sunlight into colors. This was humanity's first ever artificial creation of the rainbow, besides the one that is always naturally and passively observed when the Sun appears over the clouds while it rains, and it famed young Newton forever in the minds of many as a very bold and innovative observer of nature. What Newton showed is that multi-color light does not change its properties after being separated nor after being reflected on various objects, and that the sensation of color in our eyes is the result of the interaction between the nature of the color of the light itself and the nature of the reflecting objects themselves. This is known as Newton's theory of color.

Newton's contemporary, the English scientist Robert Hooke was one of the leaders of the Royal Society, and he considered himself to be a master in the topic of optics. When Newton published 'Of Colours', Hooke became an instant dissenter and subsequently wrote a condescending and very harsh critique of it. Newton who was then still relatively unknown, and who was seven years younger than Hooke, felt extremely disturbed by the criticism, and even more distressed by the give and take of the debate about his ideas on optics. As a result Newton began to reduce much of his social ties, and he withdrew into a sort of isolation for a while. After the death of his mother in 1679, his isolation intensified, and he became even more withdrawn for the next six years, rarely initiating exchanges on his own, and quickly breaking off intellectual exchanges which other initiated.

A much more significant and by far bitterer priority dispute with Robert Hooke occurred later with Newton's publication of Principia regarding universal gravitation. It appears that Hooke had indeed independently arrived (partially) at something close to what Newton then fully completed, yet Hooke's statements however made no mention that an inverse square law applies or might apply to these attractions. Hooke's gravitation was also not yet universal, though it approached the idea. Hooke also did not provide accompanying evidence or mathematical demonstration.

It should be noted that the years of Newton's youth were the most turbulent in the history of England. The English Civil War had begun in 1642, King Charles was beheaded in 1649, Oliver Cromwell ruled as lord protector from 1653 until he died in 1658, followed by his son Richard from 1658 to 1659, leading to the restoration of the monarchy under Charles II in 1660. How much the political turmoil of these years affected Newton and his family is unclear, but the effect on Cambridge and other universities was substantial, if only through unshackling them for a period from the control of the Anglican Catholic Church.

Newton documented his first performed alchemy experiment during 1678, having first obtained furnaces and chemicals in 1669. Experiments with metals included analysis of taste, of which there are 108 documented such undertakings, including the tasting of mercury. It is possible that as a result Newton suffered from mercury poisoning, badly affecting him physically and mentally.

Newton never married. The French writer and philosopher Voltaire, who was in London at the time of Newton's funeral, said that "Newton was never sensible to any passion, was not subject to the common frailties of mankind, nor had any commerce with women—a circumstance which was assured me by the physician and surgeon who attended him in his last moments". The widespread belief that Newton died a virgin has been commented on by several modern writers.

Chapter 32
The Publication of Principia

Early in his career, Newton was often reluctant to publish his work, despite encouragement from some of the preeminent scientists and scholars of his day. His own modesty (which didn't last forever), combined with some of the harsh criticism he received about his early discoveries, caused him to keep some ideas to himself. But eventually, with the support of Isaac Barrow (his teacher at Cambridge who also contributed earlier to calculus himself) and others, Newton began to write and publish widely.

After the informal writings of several smaller and limited versions and editions, booklets, essays, and letters to colleagues by Newton regarding his discoveries, the massive masterpiece Principia was finally published on 5 July 1687, written in Latin. The full name of the book is 'Philosophiae Naturalis Principia Mathematica' or in English as 'Mathematical Principles of Natural Philosophy'. In this work, Newton laid out the three universal laws of motion as well as the law of universal gravitation. Together, these laws describe the relationship between any object, the forces acting upon it, and the resulting motion, laying the foundation for classical mechanics. The contents of Principia contributed directly to many advances during the Industrial Revolution which soon followed. Many of these advancements continue to be the underpinnings of non-relativistic technologies in the modern world, all due to Newton's discoveries.

In the same work of Principia, Newton presented a calculus-like method of geometrical analysis of the speed of sound in air; inferred the oblateness of Earth's spheroidal figure; accounted for the precession of the equinoxes as a result of the Moon's gravitational attraction on the Earth's oblateness; initiated the gravitational study of the irregularities in the motion of the moon; provided a theory for the determination of the orbits of the comets, and much more.

Newton made clear his heliocentric view of the Solar System, although he developed it in a more modern way. This is so because already in the mid-1680 s Newton recognized the 'deviation of the Sun' from the centre of gravity of the Solar System. For Newton, it was not precisely the centre of the Sun or any other body that could

© The Editor(s) (if applicable) and The Author(s), under exclusive license to Springer
Nature Switzerland AG 2020
A. E. Kossovsky, *The Birth of Science*, Springer Praxis Books,
https://doi.org/10.1007/978-3-030-51744-1_32

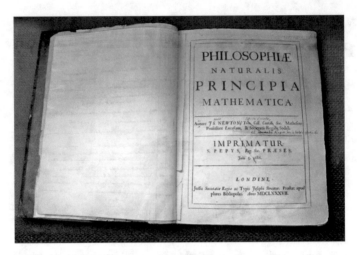

Fig. 32.1 Newton's own first edition copy of principia

be considered at rest, but rather "the common centre of gravity of the Earth, the Sun and all the Planets is to be esteemed the Centre of the World".

Figure 32.1 depicts Newton's own first edition copy of Principia, which accumulated numerous notes, corrections, and new ideas by him throughout the years, all scribed on the margins and on the opposite pages, until the publication of Principia second edition.

With the publishing of Principia, Newton became internationally recognized. He acquired a circle of admirers, including the Swiss-born mathematician Nicolas Fatio de Duillier who became the closest friend of Newton for several years. Their intense relationship however came to an abrupt and unexplained end in 1693, and at the same time Newton suffered - by his own testimony - what in more modern times would be called a severe nervous breakdown.

Despite its revolutionary content, scientists found the Principia very difficult to understand. Many of the era's scholars couldn't decipher it, including, for a time, Leibniz and Huygens, two of Newton's great contemporaries. This gap in understanding existed in part because there were few people studying mathematics at such a high level as required in reading Principia.

In the summer of 1689 Newton finally met Christiaan Huygens face to face for two extended discussions. In part because of disappointment with Huygens not being convinced by Newton's argument for universal gravity, Newton then in the early 1690s initiated a radical rewriting of the Principia. Newton had also been urged to make a new edition of the Principia since copies of the first edition had already become very rare and expensive within a few years after the publication. At last, on 30 June 1713 a second edition was published after a great deal of effort by Newton and his assisting colleagues. Though the original plan for a radical restructuring had long been abandoned, the fact that virtually every page of the Principia received some modifications in the second edition shows how carefully Newton reconsidered everything in it; and important parts were substantially rewritten not only in response to Continental criticisms, but also because of new data, including data from experiments on resistance forces carried out in London. Focused effort on the third edition of Principia began in 1723 when Newton was 80 years old, and it was subsequently published in 1726.

Chapter 33
The Bitter Dispute with Leibniz over Calculus Priority

Newton invented the calculus in the mid to late 1660s, almost a decade before Leibniz did so independently and ultimately more influentially. Newton later became involved in a long and very bitter dispute with Leibniz over priority in the development of calculus (a.k.a 'The Leibniz–Newton Calculus Controversy'). Most modern historians believe that Newton and Leibniz developed calculus independently, although with very different notations. Leibniz's notations though are recognized nowadays as much more convenient and efficient ones, and consequently they were adopted by continental European mathematicians. After 1820 or so British mathematicians relented and joined the continentals in using the same notations.

© The Editor(s) (if applicable) and The Author(s), under exclusive license to Springer Nature Switzerland AG 2020
A. E. Kossovsky, *The Birth of Science*, Springer Praxis Books,
https://doi.org/10.1007/978-3-030-51744-1_33

Chapter 34
The Scientific Split Between England and Continental Europe

Newton became a dominant figure in England almost immediately following publication of Principia, with the consequence that "Newtonianism" of one form or another had become firmly rooted in England within the first decade of the eighteenth century.

Newton's influence on the continent, however, was delayed by the strong opposition to his theory of gravity expressed by such leading figures as Christiaan Huygens and Gottfried Leibniz, both of whom saw the theory as invoking an occult or mystical power of action at a distance in the absence of Newton's having proposed a contact mechanism by means of which the forces of gravity could act.

Ill feelings between Newton and Leibniz had been developing below the surface, and finally came to a head in 1710 when John Keill of England accused Leibniz of having plagiarized calculus from Newton. Leibniz demanded redress from the Royal Society. The president of the Society at the time was Newton himself, and naturally the Society's 1712 published response was anything but redress, and indeed defended the accusation against Leibniz.

Leibniz and his colleagues on the Continent had never been comfortable with Principia and gravity, and with the calculus priority dispute brewing, this attitude turned into one of open hostility toward Newton's theory of gravity—a hostility that was matched in its blindness by the fervor of acceptance of the theory in England. The public elements of the priority dispute had the effect of expanding a schism between Newton and Leibniz into a schism between the English associated with the Royal Society and the continental group who had been working with Leibniz on calculus since the 1690s, including most notably Johann Bernoulli of Switzerland, and this schism in turn transformed into one between the conduct of science and mathematics in England versus the conduct on the Continent that persisted long after Leibniz died in 1716.

As the promise of the theory of gravity became increasingly substantiated with numerous experiments and studies all of which decisively validated it, starting in the late 1730s but especially during the 1740s and 1750s, Newton became an equally dominant figure on the continent, and "Newtonianism," though perhaps in more

© The Editor(s) (if applicable) and The Author(s), under exclusive license to Springer Nature Switzerland AG 2020
A. E. Kossovsky, *The Birth of Science*, Springer Praxis Books,
https://doi.org/10.1007/978-3-030-51744-1_34

guarded forms, flourished there as well. What physics textbooks now refer to as "Newtonian mechanics" and "Newtonian science" consists mostly of new and subsequent results and applications achieved on the European continent between 1740 and 1800.

Chapter 35
Newton's Later Years and Royal Mint Work

In January 1689, following the Glorious Revolution at the end of 1688, Newton was elected to represent Cambridge University in the Convention Parliament, which he did until January 1690. Newton moved to London to take up the post of warden of the Royal Mint [coins] in 1696, a position Newton held for the last 30 years of his life. Newton took the job seriously, exercising his power to reform the currency and to discipline counterfeiters.

Newton was made President of the Royal Society in 1703 and an associate of the French Académie des Sciences. Newton thus became a figure of imminent authority in London over the rest of his life, in ways that he had not known in his Cambridge years. Although Newton obviously had far less time available to devote to solitary research during his London years than he had in Cambridge, he did not entirely cease to be productive. The first English edition of his Opticks finally appeared in 1704. Other earlier work in mathematics began to appear in print, including a work on algebra called 'Arithmetica Universalis' published in 1707, another called 'De Analysi', and a tract on finite differences called 'Methodis differentialis' published in 1711.

Newton died in his sleep in London on 20 March 1727 at the age of 84. His contemporaries' conception of him nevertheless continued to expand and widen as a consequence of various posthumous publications in science and mathematics relating to his work and discoveries.

A. E. Kossovsky, *The Birth of Science*, Springer Praxis Books, https://doi.org/10.1007/978-3-030-51744-1_35

Chapter 36
Newton's Semi-heretical Christian Beliefs

Newton believed that mainstream Roman Catholicism, Anglicanism and Calvinism were heretical. He thought that the Holy Trinity, one of the main doctrines of orthodox Christianity, wasn't in line with the beginnings of early Christianity, and also that Jesus, while created by God, was not divine and thus not deserving of worship.

In the 1690s, Newton wrote several religious tracts dealing with the literal and symbolic interpretation of the Bible. Even though a number of authors have claimed that the work might have been an indication that Newton disputed the belief in Trinity, others assume that Newton did question it but never denied Trinity as such. The Christian doctrine of the Trinity holds that God is three consubstantial persons—the Father, the Son (Jesus Christ), and the Holy Spirit—as 'one God in three Divine Persons'. The opposing view is referred to as Nontrinitarianism.

Newton was a devout Christian, although it seems that he concluded that the Trinity dogma of God was a false doctrine and therefore refused ordination in the Anglican Church, a most unpopular decision that almost cost him his position at Cambridge University.

Indeed, when Newton was appointed Lucasian Professor of Mathematics at Cambridge in 1669, he nearly ran into a serious religious dispute. In those days, any colleague of Cambridge or Oxford was required to become an ordained Anglican priest. However, the terms of the professorship required that the holder not be active in the church (presumably so as to have more time for science). Newton argued that this stipulation should exempt him from the ordination requirement altogether, and Charles II—king of England, Scotland and Ireland—whose permission was needed, accepted this argument. Thus a conflict between Newton's religious views and Anglican orthodoxy was averted.

According to most scholars, Newton was a monotheist who believed in biblical prophecies but was Antitrinitarian. In Newton's eyes, worshiping Jesus Christ as God was idolatry and a fundamental sin. Historian Stephen D. Snobelen says of Newton: "Isaac Newton was a heretic. But he never made a public declaration of his private faith—which the orthodox would have deemed extremely radical. He hid his faith so well that scholars are still unraveling his personal beliefs."

A. E. Kossovsky, *The Birth of Science*, Springer Praxis Books,
https://doi.org/10.1007/978-3-030-51744-1_36

In an era notable for its religious intolerance, there are only very few public expressions of Newton's radical views, most notably his refusal to receive holy orders, and his refusal, on his death bed, to receive the sacrament when it was offered to him. Newton's writing in radical theology (for the most part) is material that had become public only since around mid-twentieth century. After he died, Newton's relatives immediately gathered and concealed most of his writings on religion and alchemy because they thought that making those writings public knowledge would have severely damaged his reputation. Many of Newton's personal papers only became available to scholars when they were released on microfilm in 1991.

Chapter 37
Newton's Three Laws of Motion

Newton's three laws of motion describe the relationship between an object; the forces acting upon it; and its motion in response to those forces. More precisely, the first law defines the force or rather its absence qualitatively; the second law offers a quantitative measure of the relationship between the force, the mass, and the acceleration; and the third law asserts that a single isolated force acting on only one object doesn't exist. These three laws can be summarized as follows:

First law
In an inertial frame of reference, an object either remains at rest or continues to move at a constant speed and in the same direction, unless acted upon by a force (the law of inertia).

Second law
In an inertial frame of reference the vector sum $\mathbf{F}$ of the three forces in all three dimensions acting on an object is equal to the mass $\mathbf{M}$ of that object multiplied by the vector sum $\mathbf{A}$ of the three accelerations of the object in all three dimensions: $\mathbf{F} = \mathbf{MA}$.

Third law
When one body exerts a force on a second body, the second body simultaneously exerts a force equal in magnitude and opposite in direction on the first body (the action-reaction law).

© The Editor(s) (if applicable) and The Author(s), under exclusive license to Springer   169
Nature Switzerland AG 2020
A. E. Kossovsky, *The Birth of Science*, Springer Praxis Books,
https://doi.org/10.1007/978-3-030-51744-1_37

Chapter 38
Newton's Law of Universal Gravitation

Newton's law of universal gravitation states that any particle in the universe attracts every other particle with a force (pointing along the line intersecting both particles) which is directly proportional to the product of their masses and inversely proportional to the square of the distance between their centers.

Law of Universal Gravitation: $\mathbf{F_G} = \mathbf{GM_1M_2/R^2}$.

The constant G is called the Universal Constant of Gravitation.

Its value is $\mathbf{G} = \mathbf{6.67 * 10^{-11}}$ m^3 kg^{-1} s^{-2}.

In order to measure G with experiments on Earth, it is necessary to know a priori the value of the mass of the Earth, but this value was not yet known in Newton's time. Newton was not able to provide the value of G even though he provided the correct theory. Since none of Newton's calculations could be used without knowing the value of G, Newton could only calculate the value of one force relative to the value of another force. This demonstrates Newton's brilliance and tenacity in building up his grand theory in spite of this obstacle! The universal gravitation constant G was not measured until seventy one years after Newton's death by the English natural philosopher and scientist Henry Cavendish with his Cavendish experiment, performed in 1798.

While Newton was able to formulate his law of gravity in his monumental work, he was deeply uncomfortable with the notion of "action at a distance" that his equations implied, but he became resigned to the fact that there was nothing more that he could do at the time. In 1692 he wrote: "That one body may act upon another at a distance through a vacuum without the mediation of anything else, by and through which their action and force may be conveyed from one another, is to me so great an absurdity that, I believe, no man who has in philosophic matters a competent faculty of thinking could ever fall into it."

© The Editor(s) (if applicable) and The Author(s), under exclusive license to Springer Nature Switzerland AG 2020
A. E. Kossovsky, *The Birth of Science*, Springer Praxis Books,
https://doi.org/10.1007/978-3-030-51744-1_38

More than two centuries later, in 1915, these objections were explained by Albert Einstein's Theory of General Relativity, in which gravitation is an attribute of the curved spacetime instead of being a force propagated between bodies. In Einstein's theory, the gravitational force is a fictitious concept, instead, energy and mass distort spacetime in their vicinity, and other particles move in trajectories determined by the geometry of such distorted spacetime.

Chapter 39
Deriving Kepler's 3rd Law from Newton's Laws (Optional)

Let us make three assumptions for the sake of simplicity:

(1) A small planet with low mass value orbits the much larger Sun with high mass value. This allows us—without significant error—to view the motions of the system as if the Sun were stationary.

(2) The system is isolated from other masses. This allows us to neglect any effects due to outside masses such as those of the other planets in the Solar System.

(3) Circular motion for the planet. This is a good approximation for the elliptical orbits of the planets in the Solar System which are all of relatively low eccentricity. Moreover, we are assuming that the Sun is at the very centre of the planet's circular orbit, and that there is nothing elliptical about the path. This assumption would best fit Venus, the least eccentric and most normal of all planets, with its low eccentricity value of 0.00677. This assumption would fit Mercury a bit less, being the most eccentric of all planets, with its eccentricity value of 0.21. By implication we assume constant speed throughout the circular orbit without the differentiation of Kepler's second law.

Let us now then geometrically derive the expression of the centripetal acceleration A_c. Centripetal acceleration literally means 'toward the center' or 'center seeking', and it refers to a body moving in a circular path of radius r and with a constant speed V. Figure 39.1 depicts the arrangement and analysis for an infinitesimal time interval in which the body moved a tiny bit forward.

The direction of the instantaneous velocity is shown at two points along the path. Acceleration is in the direction of the change in velocity, which points directly toward the center of rotation, namely towards the center of the circular path. This direction is shown with the vector diagram in the right panel of Fig. 39.1.

The change in velocity in vector terms ΔV, defined as $\Delta V = V_2 - V_1$ is seen to point directly toward the center of curvature. Because $A_c = \Delta V/\Delta t$, therefore the

© The Editor(s) (if applicable) and The Author(s), under exclusive license to Springer Nature Switzerland AG 2020
A. E. Kossovsky, *The Birth of Science*, Springer Praxis Books,
https://doi.org/10.1007/978-3-030-51744-1_39

Fig. 39.1 Calculation of
centripetal acceleration

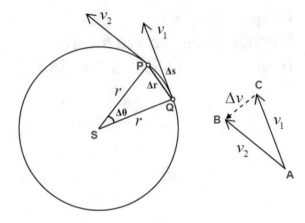

acceleration is also toward the center. Because $\Delta\theta$ is exceedingly small for infinitesimally small time differences, the round arc length Δs between points P and Q along
the circle is equal to the linear chord length Δr between points P and Q.

The direction of the centripetal acceleration is toward the center of the circle, but
what is its magnitude? In order to calculate this magnitude we first note that the
triangle formed by the velocity vectors and the triangle formed by the radii r and Δs
are similar. In addition, both triangles, SQP and ACB, are isosceles triangles with
a bit of symmetry, namely each is a triangle having two equal sides. The two equal
sides of the velocity vector triangle are the speeds $V_1 = V_2 = V$. Using the properties
of two similar triangles, we obtain:

$$\frac{\Delta v}{V} = \frac{\Delta s}{r}$$

Acceleration is $\Delta v/\Delta t$, so we first solve for Δv in the above expression and obtain:

$$\Delta V = \frac{V}{r}\Delta s$$

If we divide both sides by Δt, we get the following result:

$$\frac{\Delta V}{\Delta t} = \frac{V}{r} * \frac{\Delta s}{\Delta t}$$

Finally, since $\dfrac{\Delta V}{\Delta t} = A_C$ and $\dfrac{\Delta s}{\Delta t} = V$, namely the linear or tangential speed,
hence the concise expression of the magnitude of the centripetal acceleration is
$A_C = \dfrac{V^2}{r}$.

Applying Newton's 2nd law to an accelerating planet yields:

$$F = MA = M_{Planet} A_C = M_{Planet} \frac{V^2}{r}$$

where M_{Planet} is the mass of the moving planet, V is its speed, and r is the radius of its circular orbit (i.e. distance from the Sun). The centripetal force in this case is provided by gravity; hence Newton's universal gravitation can also be applied, so that:

$$F_G = \frac{G M_{Sun} M_{Planet}}{r^2}$$

Equating the two expressions above for the force, we obtain:

$$\frac{G M_{Sun} M_{Planet}}{r^2} = \frac{M_{Planet} V^2}{r}$$

Cancelling out r and M_{Planet} we get:

$$\frac{G M_{Sun}}{r} = V^2$$

But the planetary speed V is given by the distance the planet moves in one revolution [i.e. the circumference of the circular path] divided by the time it takes to do it [time period for one full revolution around the Sun], hence:

$$V = \frac{2\pi r}{T}$$

Substituting this expression for V in the earlier equation, we get:

$$\frac{G M_{Sun}}{r} = V^2 = \left(\frac{2\pi r}{T}\right)^2$$

Re-arranging the exponentiations, we get:

$$\frac{G M_{Sun}}{r} = \frac{4\pi^2 r^2}{T^2}$$

Solving for T^2 we get:

$$\frac{T^2}{1} = \frac{4\pi^2 r^3}{G M_{Sun}}$$

Utilizing the letter D for the *Distance* of the planet from the Sun, instead of the letter r for the *radius* of the planetary path (which is the same concept), we finally arrive at Kepler's 3rd Law:

$$\mathbf{T}^2 = \frac{4\pi^2}{GM_{Sun}}\mathbf{D}^3$$

This is simply Kepler's 3rd law as in the form shown in Fig. 12.2; where the fraction $4\pi^2/GM_{Sun}$ can be considered as the constant K; and thus can be written as:

(Time of one Full Revolution)^2 = K * (Distance from Sun)^3.

Chapter 40
Newton's Generic Laws Versus Kepler's Specific Laws

In a similar (mathematical) fashion, Kepler's 1st law, namely the possibility of an elliptical path for a body under the influence of gravity, is shown compatible with Newtonian Mechanics. Kepler's 2nd law of differentiated speeds along an elliptical path is also readily derived from Newtonian Mechanics.

But young Newton at the age of about 25 years had to work all this *backward*! Namely, from knowing Kepler's laws to inventing the theoretical basis for it!

Kepler's laws are wonderful as a description of the motions of the planets. However, they provide no explanation of why the planets move in this way. Moreover, Kepler's 3rd law works smoothly for all the planets around the Sun with a singular value of K, but for the moons of Jupiter it requires a totally different value for the constant K; and it does not readily apply for the Moon's orbit around the Earth. Newton on the other hand provided a more general explanations and descriptions of the motions of the planets, the Moon, the moons of Jupiter, and any other celestial body, in one fell swoop, using a singular and consistent set of constants and equations.

By showing how Kepler's laws could be derived from his theory, as well as explaining projectile motion and freely falling bodies as described by Galileo, all with the same theory, Newton united the heaven and the Earth. Before Newton, the heavens were thought to work according to rules quite different from the ones which governed things on Earth. [Note: For Kepler's 3rd law, constant K for Jupiter's moons is indeed of larger value than constant K for the planets revolving around the Sun, and this is explained neatly and exactly by Newtonian Mechanics, because it is the lighter mass of Jupiter, instead of the heavier mass of the Sun, that is inserted in the nearly final formula of the previous chapter, as $(4\text{pi}^2)/(GM_{\text{Jupiter}})$ for the generic expression of constant K.]

A. E. Kossovsky, *The Birth of Science*, Springer Praxis Books, https://doi.org/10.1007/978-3-030-51744-1_40

Chapter 41
The Rationale Behind the Gravitational Formula

Let us justify the part of Newton's expression which accounts for the masses of the two objects within the overall formula of universal gravitational law.

$$F = \frac{G*(M_1*M_2)}{R^2}$$

The confluence of the magnitudes of the two masses on resultant gravitational force is expressed via the *multiplication* of their masses, and this *product* in itself is a factor in the force.

Certainly we need an expression that correlates positively with M_1 as well as one that correlates positively with M_2. Intuitively, the larger the value of M_1 the stronger the force; the larger the value of M_2 the stronger the force; and of course the larger the values of both M_1 and M_2 the stronger the force. These correlative facts could manifest themselves arithmetically within the expression of gravitational force via the operation of addition $(M_1 + M_2)$ for example, or via the operation of multiplication (M_1*M_2) for example—as Newton chose, or possibly via the involvement of some powers, such as for example $(M_1^P + M_2^P)$ or $(M_1^P * M_2^P)$, where exponent P is greater than 1.

To recap: the more (overall) mass we have in the system the stronger the gravitational force. The dilemma is how best to account arithmetically for this obvious fact in the expression of universal gravitation.

For two masses with a fixed distance between them, one can view Newton's expression for the gravitational force as a function of just two variables, namely M_1 and M_2, with G/R^2 written more succinctly as a constant K, as follows:

$$F(M_1, M_2) = K * (M_1 * M_2)$$

Let us suggest some reasonable alternatives to Newton's insistence on a simple product; namely, alternatives where both M_1 and M_2 are treated equally; and where the above-mentioned correlative principles still hold true.

© The Editor(s) (if applicable) and The Author(s), under exclusive license to Springer Nature Switzerland AG 2020
A. E. Kossovsky, *The Birth of Science*, Springer Praxis Books,
https://doi.org/10.1007/978-3-030-51744-1_41

Here are some possibilities—where constant K is left flexible, possibly assuming any other 'experimentally-verified' value:

$$F(M_1, M_2) = K * 7(M_1 * M_2)$$
$$F(M_1, M_2) = K * 4(M_1 * M_2)$$
$$F(M_1, M_2) = K * (M_1 + M_2)$$
$$F(M_1, M_2) = K * (M_1 + M_2)^3$$
$$F(M_1, M_2) = K * (M_1 * M_2)^2$$
$$F(M_1, M_2) = K * (M_1^5 * M_2^5)$$
$$F(M_1, M_2) = K * (M_1^6 + M_2^6)$$
$$F(M_1, M_2) = K * (M_1^{M2} + M_2^{M1})$$
$$F(M_1, M_2) = K * (M_1^{M2} * M_2^{M1})$$

In principle, it should be possible to verify the correctness of Newton's choice empirically by performing some experiments with several rocks of varying weights and observe that 'it works'.

On the other hand, one could substantiate Newton's (quantitative) assertion conceptually (or rather qualitatively) by applying thought experiments instead of physical ones.

What's wrong with all the expressions for the gravitational force above? The answer is that none of these expressions is capable of accounting for the aggregation of mini forces acting between two objects—where each object consists of numerous 'basic' or 'atomic' units of mass. Only Newton's expression represents the correct aggregated gravitational force in the interactions between two individual objects containing multiple mass-points.

The net or macro gravitational force between two solid rocks is simply the aggregation of numerous (basic) mini forces between the micro constituents [say protons and neutrons] located within the 1st rock and the micro constituents [say protons and neutrons] located within the 2nd rock. Those protons and neutrons are firmly imbedded within each rock as the electromagnetic force easily overwhelms the gravitational force, so individually these constituents are not allowed to move within either rock because of any supposed weak gravitational forces, only the rocks themselves are allowed to move as whole, empowered by the aggregation of these mini gravitational forces. It is this macro movement and this aggregated force acting on the rocks that Newton's formula refers to (as opposed to the micro gravitational forces between the basic protons or neutrons).

Let us give five numerical examples of the aggregation of micro forces into an overall macro force.

In Fig. 41.1, $M_1 = 3$ and $M_2 = 2$, so there are 6 such basic forces to add. This is consistent with $F = K*(M_1*M_2) = K*3*2 = K*6$.

In Fig. 41.2, $M_1 = 4$ and $M_2 = 1$, so there are 4 such basic forces to add. This is consistent with $F = K*M_1*M_2 = K*4*1 = K*4$.

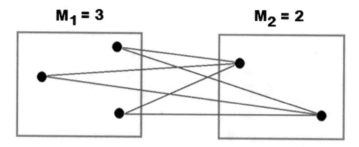

Fig. 41.1 Macro gravitational force for 3 + 2 micro elements

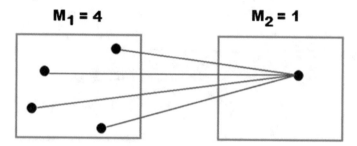

Fig. 41.2 Macro gravitational force for 4 + 1 micro elements

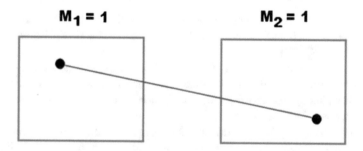

Fig. 41.3 Macro gravitational force for 1 + 1 micro elements

In Fig. 41.3, $M_1 = 1$ and $M_2 = 1$, so there is only 1 such basic forces "to add". This is consistent with $F = K*M_1*M_2 = K*1*1 = K*1 = K$.

This **is** the basic gravitational force in its purest form—the force whose aggregation drives the macro movement of the rocks.

In Fig. 41.4, $M_1 = 5$ and $M_2 = 0$, so there exist no such atomic forces whatsoever to add. This is consistent with: $F = K*M_1*M_2 = K*5*0 = 0$. (The 2nd rock is just a shell—an empty form.)

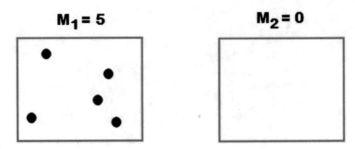

Fig. 41.4 Macro gravitational force for 5 + 0 micro elements

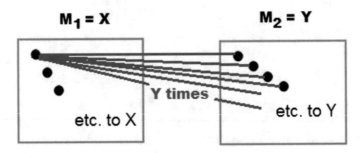

Fig. 41.5 Macro gravitational force for X + Y micro elements

Figure 41.5 depicts the generic case where $M_1 = X$ and $M_2 = Y$, therefore there are 'Y times X' such basic forces to add, namely X*Y. This is consistent with $F = K*(M_1*M_2) = K*X*Y$.

In summary: The expression $K*(M_1*M_2)$ is simply an accounting scheme designed to aggregate all the gravitational forces occurring at the micro level. None of the other alternatives above for the expression of the gravitational force could properly account for such aggregation.

For a rock weighing 1 kg in gravitational interaction with another rock weighing 1 kg, the system contains enormous basic forces at the sub-atomic level. Since the proton weighs $1.6727*10^{-27}$ kg, and the neutron weighs $1.6750*10^{-27}$ kg (ignoring the electron which weighs $9.110*10^{-31}$ kg), then each one kilogram of mass—of whatever element or molecule type and whatsoever mix—contains approximately $(1)/(1.6738*10^{-27})$ or $5.97425*10^{+26}$ basic protons or neutrons at the micro level.

Here, there are $(5.97425*10^{+26})*(5.97425*10^{+26})$ or simply $3.56917*10^{+53}$ basic or micro gravitational interactions between these two 1-kg rocks.

Interestingly, the most balanced and fair allocation of fixed amount of mass between two boxes yields the strongest gravitational force between them.

Let us consider the case of a two-object gravitational interaction, having fixed amount of overall mass within the entire system, say 10 kg, namely fixed-sum or constant-sum situations. If mass-allocation is assumed to be totally flexible so that we can draw any amount of mass from one box and place it in the other box, and if distance is fixed at some particular value and thus it is not a factor in resultant gravitational force, then the question that arises naturally here is as follows: under what arrangement of mass allocations do we obtain the maximum gravitational force?

If we place the entire mass of 10 kg in one box, and nothing in the other box, the factor is then $0*10 = 0$, so that there exists no gravitational force whatsoever. If we place 1 kg in one box, and 9 kg in the other box, the factor is then $1*9 = 9$. If we place 2 kg in one box, and 8 kg in the other box, the factor is then $2*8 = 16$, but still, we have not reached the optimum force here. If we place 3 kg in one box, and 7 kg in the other box, the factor is then $3*7 = 21$, and which is even bigger. If we place 4 kg in one box, and 6 kg in the other box, the factor is then $4*6 = 24$.

Clearly, by keeping the weights of the two boxes as even and as fair as possible; dividing overall mass equally; we obtain the maximum possible product of the masses and thus the maximum possible gravitational force. Hence the arrangement of 5 kg in one box, and another 5 kg in the other box, yields the factor of $5*5 = 25$, and results in the maximum possible gravitational force. In sharp contrast, extreme inequality where one empty box contains zero mass and another box contains the entire overall mass yields zero gravitational force.

Chapter 42
Rationale of Newton's Formula Depends on Relative Distances (Optional)

Such atomic and detailed vista of Newton's expression assumes that distances between the micro protons and neutrons within each rock are negligible as compared with the much larger distance between the macro rocks themselves, so that, for all practical purposes and with high degree of approximation, all the basic forces between the two rocks are of the same distance approximately, and that therefore a singular value for the radius r can be used in the expression for the overall force. On the other hand, if the distance between the rocks is of similar dimension as the distances between the micro protons and neutrons within each rock, then most of the basic forces between the two rocks are of distinct distances, and therefore a wide variety of values for the radius r must be used in the expression for the overall force, making calculations extremely complex, and this requirement totally ruins the basis for the use of Newton's simple and unified expression $F = K*(M_1*M_2)$.

Figure 42.1 depicts the case of atomic and rocks distances being of comparable dimensions and where the usage of Newton's expression appears invalid. The longer basic force in the figure is deserving of $1/8^2$ or 0.0156 distance factor; while the shorter basic force in the figure is deserving of $1/3^2$ or 0.1111 distance factor, and which is 7 times as large as the former factor. Such huge disparities in relative distances appear to invalidate the entire accounting scheme.

Figure 42.2 depicts the case of tiny atomic micro distances as compared with the much larger distance between the macro rocks, and where the usage of Newton's expression is certainly valid. Figure 42.2 is not drawn exactly to scale for lack of space, as the two lines should extend much farther. The slightly longer basic force in the figure is deserving of $1/908^2$ or 0.000001213 distance factor; while the slightly shorter basic force in the figure is deserving of $1/903^2$ or 0.000001226 distance factor, and which is merely 1% larger than the former factor. Such tiny disparities in relative distances ensure that the entire accounting scheme is correct to very high degree of approximation.

In reality, the formal definition of Newton's law of universal gravitation states that every *point mass* attracts every single other *point mass* by a force pointing along the line intersecting both points—and with a magnitude according to the standard

© The Editor(s) (if applicable) and The Author(s), under exclusive license to Springer Nature Switzerland AG 2020
A. E. Kossovsky, *The Birth of Science*, Springer Praxis Books,
https://doi.org/10.1007/978-3-030-51744-1_42

Fig. 42.1 Comparable
macro and micro distances

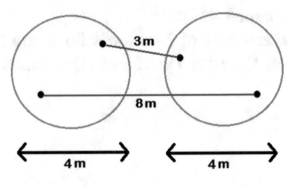

Fig. 42.2 Larger macro
distances and smaller micro
ones

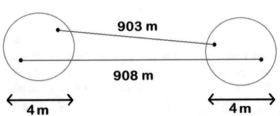

formula. If the bodies in question have *spatial extent* instead of being an abstractly
theorized point masses—as all real-world objects do—then the gravitational force
between them is calculated by summing the contributions of the hypothetical point
masses which constitute the bodies. Ignoring the atomic and molecular theory in
chemistry, and assuming continuous and smooth material, then in the limit, as the
component point masses become "infinitely small", this entails applying Calculus
and integrating the forces (in vector form) over the extents of the two bodies.

Fortunately, in spite of the apparent need to apply a wide variety of distinct r
distances to objects with spatial extent as in Fig. 42.1, Newton's formula is nonetheless often protected by arguments of symmetry, so that a singular average value of r
can be applied. Indeed Newton himself argued as such.

In classical mechanics, the *Shell Theorem* gives gravitational simplifications that
can be applied to objects inside or outside a spherically symmetrical body, having
uniform density of mass so that its entire mass is dispersed equally and smoothly
all throughout the sphere or the shell. This theorem has particular application to
astronomy since almost all celestial objects are nearly or approximately spherical in
form. Newton himself proved the shell theorem and stated that:

(1)   A spherically symmetric body affects external objects gravitationally as though
      all of its mass were concentrated at a point at its centre. Figure 42.3 depicts the
      arrangement.
(2)   If the body is a spherically symmetric shell (i.e., a hollow ball), no net gravita-
      tional force is exerted by the shell on any object inside, regardless of the object's
      location within the shell. Figure 42.4 depicts the arrangement.

Fig. 42.3 Sum of forces
outside the sphere reduced to
a point

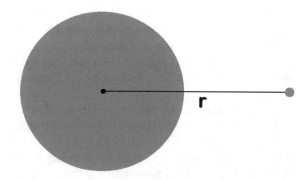

Fig. 42.4 Zero gravitational
force inside symmetrical
shell

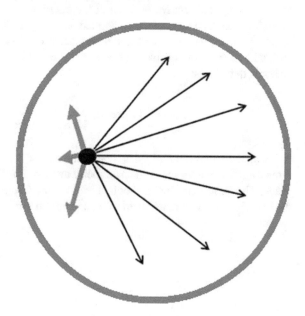

These results were important to Newton's analysis of planetary motion. While
they are not immediately obvious, they can be proven with Calculus, and arrived via
the aggregation of all the tiny forces of attraction between the infinitesimally small
parts of the spherical body and the external or internal object.

Moreover, in Prinicipia, Proposition LXXV, Theorem XXXV, Newton demonstrated that for two spherical masses, the overall gravitational force operates as if
confined to their centers; so that, one sphere will act upon another sphere, with a force
directly proportional to the product of the values of the masses of the two spheres,
and inversely proportional to the square of the distance between their centers. The
immediate implication here is that the attraction force of the Sun on planet Mars for
example, can be thought to be conducted simply via their centers.

Hence if Fig. 41.5 relates to two perfect spheres with smooth and uniform atomic
densities, then all the various micro forces having a variety of r distances between

the atomic particles can be averaged out in one fell swoop via the factor $1/R^2$, where R is simply the distance between their centers—and the application of Newton's F = $K*(M_1*M_2)$ format is wholly justified.

The above discussion regarding averaging out distinct micro gravitational forces should not be confused with the concepts of Center of Mass, Center of Gravity, or the Centroid.

Centre of Mass is the point at which the distribution of mass is equal in all directions, and it does not depend on the (possibly differentiated) gravitational field. Centre of Gravity is the point at which the distribution of weight is equal in all directions, and it does depend on the (possibly differentiated) gravitational field. Centre of Mass is defined conceptually as the average position of all the mass points, or as the weighted relative position of the distributed mass. Center of Mass is also the point such that when a force is applied onto it by a pin or a stick, the entire object moves in the direction of the force without rotating. For an object with N particles, each with M_i quantity of mass at positional vector V_i, the Center of Mass positional vector is defined as:

$$\text{Center of Mass} = \frac{1}{M_{TOTAL}} * \sum_{i=1}^{N} M_i * V_i$$

In mathematics and physics, the Centroid or 'geometric center' of a two-dimensional plane figure is the arithmetic average position of all the points in the shape. The geometric Centroid coincides with the Center of Mass of a flat object if the mass is uniformly distributed over the entire figure.

Chapter 43
The Observer, the Organizer, and the Theorist

The fascinating story of Tycho, Kepler, and Newton, illustrates humanity's profound challenge in fully understanding and describing the Solar System in terms of the correct heliocentric model, and ultimately in terms of the physical laws of motion, and the gravitational forces applied to the planets. The triumph of discovery here was made due to the distinct contributions of three very different types of scientists, each bringing into the inquiry unique approach, focus, and temperament.

The focus in this chapter is solely on the direct contribution of Kepler's planetary laws, arrived via Tycho's astronomical data, to the theorizing and creative mind of young Newton during the Great Plague, resting and contemplating the Solar System among other things. This chapter (unfairly perhaps) downplays Galileo's indirect terrestrial contributions to the understanding of the Solar System. Galileo would raise abstract and generic questions about terrestrial motion, acceleration, and inertia. Then, in order to answer these questions he would first elaborate on these issues mathematically and geometrically. And finally he would design specific physical instruments for experiments and measurements in order to empirically test hypotheses. Galileo did not focus on the mechanism and numerical measures of the Solar System, except for his mysterious and puzzling hint at celestial application of his terrestrial work. We shall temporarily then focus exclusively here on Tycho, Kepler and Newton.

Tycho was the meticulous observer. Kepler was the organizer, or rather the data analyst. And Newton was the natural philosopher and the theorist. Each played a crucial yet very distinct role in the entire process of discovery. All in all, they worked in unison, as if in symphony, yet their accomplishments occurred sequentially in different periods and times. The contribution of each one would have been meaningless without the contributions of the other two. The work of only one of them, or only two of them, would not have led to any fruitful discoveries whatsoever. The combined effort of these three very different personalities was a revolution in our understanding of the Solar System, and it also led to the discovery of the laws of physics in general.

© The Editor(s) (if applicable) and The Author(s), under exclusive license to Springer Nature Switzerland AG 2020
A. E. Kossovsky, *The Birth of Science*, Springer Praxis Books,
https://doi.org/10.1007/978-3-030-51744-1_43

Tycho Brahe was a dedicated and enthusiastic observer. He took enormous amount of quite precise data (relative to the standards of his era) on the positions and timing of the planets as observed by naked eyes, without the aid of any telescope. He then carefully recorded all his observations.

Tycho himself didn't believe in the heliocentric model even though his work led directly to it; rather he believed in a modified or hybrid model which fused together the heliocentric and geocentric models. Fortunately for science and humanity, Tycho was an honest scientist, and he did not fudge his data or cheated in any way in order to prove his misguided theory.

Kepler's contribution was made via his thorough data analysis and attempts at summarization of Tycho's data. Kepler took Tycho's data and through a lot of computational and organizational work, was able to fit the entirety of Tycho's planetary data with three simple laws. Kepler's work also decisively supported the simple heliocentric model of the Solar System.

Kepler condensed Tycho's data, and let the planets be described via few variables; namely eccentricity of the elliptical path, as well as either the distance from the Sun, or alternatively the time for one complete revolution. Using Kepler's laws one could predict positions and timing of planets at any time in the future or in the past.

Kepler's laws are elegant, concise, and useful, yet they are still merely empirical laws, lacking any underlying fundamental physics to explain and justify them. This is why Kepler is called the (data) organizer.

The third personality in this drama is Newton, who stated his three laws of motion and universal gravitation in extreme generality, without referring specifically to any moving planet, falling apple, or cannon ball trajectory. Clearly, Newton was directly and crucially influenced and aided by Kepler's laws in his discoveries. Moreover, Newton was then able to derive and explain Kepler's three laws from his general physical theory.

What if there was no Kepler, or that he would have never been invited to Prague by Tycho? Could Newton have discovered universal gravitation and his three laws of motion directly from the (huge and messy) raw data of Tycho, and from the terrestrial hints coming from Galileo? Almost certainly not! By organizing and summarizing Tycho's data, Kepler gave Newton the general vista of planetary motion and the basis upon which to build his physics. And now, once Newton made his discoveries, there is really no need to verify Newton's work by referring to the entirety of the data recorded by Tycho. Since Newton's theoretical laws correspond to Kepler's organizational laws, and since Kepler's organizational laws correspond to Tycho empirical data, hence by implication Newton's theoretical laws correspond to Tycho's empirical data as well.

After Newton, the Copernican Revolution had triumphed supreme, since now there was a physical theory explaining how and why things are the way they are. The motions, timings, speeds, and shapes of the planets' orbits were all derived, determined, predicted, and confirmed, by Newton's laws. It was finally understood that the planets are moving under the perfectly balanced tug-of-war between the forward tendency of inertia and the gravitational pull of the Sun. This was by far more decisive confirmation of the heliocentric model than merely arguments of simplicity and parsimony. At long last, the geocentric model was completely discredited.

Tycho the observer, Kepler the organizer, and Newton the theorist, will forever be remembered as an extraordinary team who sequentially led to the correct understanding of the Solar System. They will also be remembered as three in a team of six midwives attending the (long, difficult, and at times painful) birth of science.

The entire multi-generational intricate and tangled process of discovery can be summarized and expressed with the phrase:

'Data to equations to theory'.

Chapter 44
The European Nationalities of the Six Scholars

The six extraordinary scholars who ignited the Scientific Revolution represent, to a great extent, the diversity of the nationalities on the European continent.

Nicolaus Copernicus—Polish
Tycho Brahe—Danish/Swedish
Johannes Kepler—German
Galilei Galileo—Italian
René Descartes—French
Isaac Newton—English

Their common mode of communication was Latin.

With the exclusion of Australia, Europe is the smallest of all the continents, and distances between nations are not really that great. Inter-regional communication and transportation were relatively well-developed at that era of the Renaissance, due mostly to the earlier beneficial influence of the Roman Empire which politically united much of the territory at one point in time; established common standards; common language for scholars (Latin); and a common religion (almost). The Roman Empire also built many of the roads around Europe, not only those leading to Rome, but also peripheral ones.

Figure 44.1 depicts the nearly even and wide distribution of these six scholars throughout Europe. The initial letter of the last name is used to indicate a given scholar, so "**N**" for Newton, "**K**" for Kepler, and so forth.

Not to be overlooked in this context is the country **Greece**, slightly to the southeast of the concentration of these six letters; and in spite of it earning no letters whatsoever on this particular map of Fig. 44.1. Surely Greece deserves to be littered with numerous letters representing mathematicians, philosophers, and quasi-scientists from around two millenniums before the Scientific Revolution, as these ancient personalities laid the broad foundation for all that was consequently discovered later.

In addition, the dormant intellectual prowess of Slavic **Russia** which awoke shortly after Newton's era would certainly deserve several letters on this map as well. Most notably are the mathematician Nikolay Lobachevsky (1792–1856) known as

© The Editor(s) (if applicable) and The Author(s), under exclusive license to Springer Nature Switzerland AG 2020
A. E. Kossovsky, *The Birth of Science*, Springer Praxis Books,
https://doi.org/10.1007/978-3-030-51744-1_44

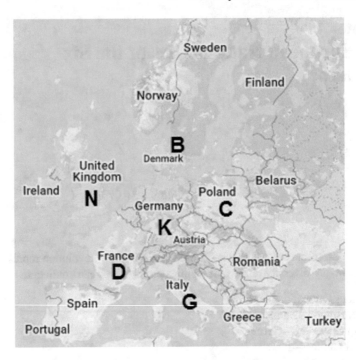

Fig. 44.1 The distribution of the six scholars in Europe

the founder of Hyperbolic Geometry which was later recognized as a valid alternative to Euclidean Geometry; the chemist Dmitri Mendeleev (1834–1907) who formulated the Periodic Law and created his own version of the Periodic Table of the elements; the Russian polymath Mikhail Lomonosov (1711–1765) who discovered the law of conservation of mass in chemical reactions; and the mathematician Pafnuty Cheby-shev (1821–1894) well-known for his work in the fields of probability, statistics, and number theory, and for his formulation of the Chebyshev Inequality.

The Scientific Revolution appears as a continent-wide project spanning several generations. However, as was noted in the first chapter, Europe at that era was not a prosperous continent, but rather one torn by senseless wars, divided violently between the newly formed Lutheran Church and the older Catholic Church, and often hit by local outbreaks of the Great Plague, until it started to subside to some extent around 1667. Indeed, one of the last significant recurrences on the continent was the Great Plague of London in 1665 which killed up to 100,000 people, but spared our young Newton.

Chapter 45
The Supposed Vocations of the Six Scholars

In their youth, none of the six scholars initially focused on the study of mathematics and science as their chosen vocation, rather they all had a very different idea of which discipline they wanted to study and what they wished to become, or in some cases that was what their practical or overbearing parents thought they should become:

Nicolaus Copernicus—Christian Canon Law
Tycho Brahe—Lawyer of the Royal Court
Johannes Kepler—Religion, wished to become a minister
Galilei Galileo—Medical doctor
René Descartes—Civil Law, becoming a lawyer
Isaac Newton—Farmer working in agriculture!

Luckily for science, they either disobeyed their parents by studying mathematics, science, and philosophy, instead of the more profitable professions with guaranteed salaries, or they themselves changed their minds about what they wanted to study.

© The Editor(s) (if applicable) and The Author(s), under exclusive license to Springer Nature Switzerland AG 2020
A. E. Kossovsky, *The Birth of Science*, Springer Praxis Books,
https://doi.org/10.1007/978-3-030-51744-1_45

Chapter 46
A Tale of Two Laws—Bode's Law and Kepler's Law

Johann Elert Bode (1747–1826) was a German astronomer who lived about two centuries after Kepler. Bode attempted anew to answer the same question that initially concerned Kepler before his discoveries of the three planetary laws, namely "why are the planets spaced around the Sun as they are?" Kepler's first answer was his failed constructions of inscribed and circumscribed circles around regular polygons, followed by his second answer of the constructions of inscribed and circumscribed spheres around Platonic solids.

In 1768 Bode published his novel idea expressed as a planetary law which attempted to explain the particular values of the distances of the planets from the Sun via a concise mathematical formula, and without any reference to geometrical constructs.

In fact, such a law was first proposed two years earlier by another German astronomer named Johann Daniel Titius (1729–1796) in 1766. Although the law is now often formally called the Titius—Bode law, it is also more commonly called Bode's law. Subsequently, Bode gave Titius clear recognition of his priority.

The focus here is merely on the planetary distances from the Sun, not on the **dynamic** variables of the time of orbit, motion, or speed. One might describe Bode's hypothesis as an analysis on a **static** system of objects (the planets) in relative position from a fixed point (the Sun). One recalls Kepler's failed hypothesis regarding the five Platonic symmetrical solids in his attempt to explain the distances of the six known planets of his era. Here Bode proposed a more flexible hypothesis which could account for any number of planets, not merely the rigid six-planet system of Kepler's Platonic Solids.

The astronomical community gradually began to show interest in Bode's hypothesis because it potentially served also as a predictor of other possible planets (yet unknown) which might exist out there in the Solar System. In other words, Bode's formulation had the supposed ability to point to exact distances (sky locations) to be searched, in the hope of finding new planets beyond the six known ones of that era.

© The Editor(s) (if applicable) and The Author(s), under exclusive license to Springer Nature Switzerland AG 2020
A. E. Kossovsky, *The Birth of Science*, Springer Praxis Books, https://doi.org/10.1007/978-3-030-51744-1_46

Fig. 46.1 Planetary
distances in au/10

PLANET	DISTANCE
Mercury	3.9
Venus	7.2
Earth	10.0
Mars	15.2
Jupiter	52.0
Saturn	95.8

Indeed, in 1781, just several years after Bode's publication, Uranus was discovered and was perfectly fitting Bode's formulations, lending Bode's law strong support and enthusiasm.

Let us take a close look at Bode's formulation.

Bode rescaled planetary data to be expressed in units of (1/10)*au, namely 0.1 of astronomical unit. The astronomical unit **au** is a unit of length equal to the distance from the Earth to the Sun, and which is about 150 Mkm, or more precisely 149, 597, 870.7 km. In summary, original au—based planetary data was multiplied by a factor of 10 for Bode's analysis. Figure 46.1 depicts distances using this au/10 scale for the six known planets at the time of Bode's Publication. The definition of distance here refers to the orbit's semi-major axis, namely the average between Perihelion and Aphelion.

The mathematical expression Bode attempted to fit those distances is:

$$\text{Distance} = 4 + 3 * 2^n$$

$$n = \{-\infty, 0, 1, 2, 3, \text{etc.}\}$$

More concisely this can be written as:

$$\text{Distance} = 4 + 3 * 2^{\{-\infty,0,1,2,3,\text{etc.}\}}$$

Planetary distances are therefore the sequence:

$$4 + 3 * 2^{-\infty}, 4 + 3 * 2^0, 4 + 3 * 2^1, 4 + 3 * 2^2, 4 + 3 * 2^3, \text{etc.}$$

Admittedly, the start at negative infinity (i.e. $-\infty$) for n of Mercury is rather odd and may seem arbitrary, but beyond Mercury the value of n increases steadily and consistently by one integer at a time.

It is noted that $2^{-\infty} = 1/2^{+\infty} = 1/\infty = 0$, hence $4 + 3 * 2^{-\infty} = 4 + 3 * 0 = 4$.

Table in Fig. 46.2 calculates those values and compared them with actual planetary distances.

Fig. 46.2 Bode's formula
and possible planetary fitness

PLANET	ACTUAL	IDEA	IDEA IS
Mercury	3.9	$4 + 3*2^{-\infty}$	4.0
Venus	7.2	$4 + 3*2^0$	7.0
Earth	10.0	$4 + 3*2^1$	10.0
Mars	15.2	$4 + 3*2^2$	16.0
???	???	$4 + 3*2^3$	28.0
Jupiter	52.0	$4 + 3*2^4$	52.0
Saturn	95.8	$4 + 3*2^5$	100.0

The idea is excellent up to and including Mars, but it fails to work after Mars, unless we assume that there exists a hidden planet between Mars and Jupiter, and in which case there is an excellent fit here overall, including good fit for Jupiter and Saturn. Hence, as originally published, the law was approximately satisfied by all known planets, Mercury through Saturn, but with a glaring gap between the fourth and fifth planets.

Except for the additive term 4 which becomes quite insignificant after several terms, Bodes's law ultimately points to an exponential 100% growth series for the distances of the outer planets, where terms are doubling approximately with each new sequence.

The law was initially regarded merely as interesting, but of no great importance, until the discovery of Uranus in 1781 by William Herschel in England. Uranus' semi-major axis was measured at 19.22 au, and this value happened to fit nicely into Bode's sequence which predicted 19.60 au for the distance of the 8th planet. In a March 1782 treatise, Bode proposed the name 'Uranus' for the newly discovered planet, which is the Latinized version of the Greek god of the sky Ouranos, and indeed this name was eventually adopted.

Based on the discovery of Uranus, Bode then urged a search for a fifth planet between Mars and Jupiter. He wrote "Can one believe that the Founder of the universe has left this space empty?" Bode's call culminated in a group formed for this purpose, the so—called "Celestial Police" in order to search the sky for the supposed planet. However, before the group initiated any search, they were trumped by the discovery of the asteroid Ceres in 1801 by Giuseppe Piazzi, from Palermo Italy, at Bode's predicted position. Ceres is the largest object in the Asteroid Belt, circling the sun at 2.77 au distance, and this actual position was found to be almost exactly at Bode's predicted position of 2.80 au for the fifth planet. Consequently, Bode's law was widely accepted and it gained considerable fame.

Table in Fig. 46.3 calculates Bode's predictions together with the actual planetary distances after the discoveries of Ceres and Uranus. The chart in Fig. 46.4 demonstrates the great success of Bode's law in fitting theoretical and actual values after the discoveries of Ceres and Uranus.

Fig. 46.3 Bode's formula and planetary distances after 1801

PLANET	ACTUAL	IDEA	IDEA IS
Mercury	3.9	$4 + 3*2^{-\infty}$	4.0
Venus	7.2	$4 + 3*2^0$	7.0
Earth	10.0	$4 + 3*2^1$	10.0
Mars	15.2	$4 + 3*2^2$	16.0
Ceres	27.7	$4 + 3*2^3$	28.0
Jupiter	52.0	$4 + 3*2^4$	52.0
Saturn	95.8	$4 + 3*2^5$	100.0
Uranus	192.2	$4 + 3*2^6$	196.0

Fig. 46.4 Chart of good fitness of Bode's law after 1801

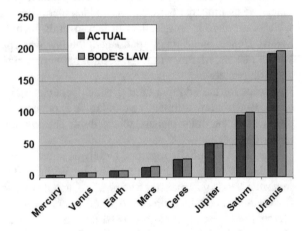

It works! What a beautiful fit!

Suddenly bad news strikes!

In 1846 Neptune was discovered, and it did not fit Bode's law.

Neptune's semi-major axis was measured at 30.11 au, and this value is starkly different than Bode's prediction of 38.80 au.

The discovery of Pluto in 1930 confounded the issue still further. Pluto's semi-major axis was measured at 39.54 au. Although nowhere near its supposed 77.20 au position as predicted by Bode's law, it was roughly at the position the law had predicted for Neptune, namely 38.80 au.

Moreover, the planetary status of Pluto came into doubt in the following years, further confounding matters. It is simply not a straightforward matter to assign ordinal numbers for the values of n in Bode's formula to all of these objects in the Solar System, such as proper planets, smaller planet—like objects, asteroids, big rocks, and all the other types of lesser objects.

Fig. 46.5 Bode's formula
and planetary distances after
1930

PLANET	ACTUAL	IDEA	IDEA IS
Mercury	3.9	$4 + 3*2^{-\infty}$	4.0
Venus	7.2	$4 + 3*2^0$	7.0
Earth	10.0	$4 + 3*2^1$	10.0
Mars	15.2	$4 + 3*2^2$	16.0
Ceres	27.7	$4 + 3*2^3$	28.0
Jupiter	52.0	$4 + 3*2^4$	52.0
Saturn	95.8	$4 + 3*2^5$	100.0
Uranus	192.2	$4 + 3*2^6$	196.0
Neptune	301.1	$4 + 3*2^7$	388.0
Pluto	395.4	$4 + 3*2^8$	772.0

Fig. 46.6 Chart of the
discrepancy in Bode's law
after 1930

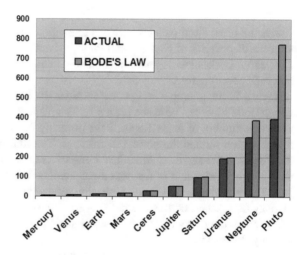

In addition, the large number of asteroids discovered in the belt beyond Mars potentially removed Ceres from the list of planets.

The subsequent discovery of the Kuiper belt, and in particular the discovery of the object Eris, which is more massive than Pluto yet does not fit Bode's law at all, further discredited the formula.

Bode's law falls apart!

The rise and fall of Bode's law in astronomy is sometimes referred to as an example of **fallacious reasoning**.

Table in Fig. 46.5 calculates Bode's distance predictions together with the actual distances of the (supposed) planets after the discoveries of Neptune and Pluto. The table is based on two assumptions:

(1) That Ceres or its associated Asteroid Belt can be considered a proper planet, and (2) that Pluto can be considered a proper planet. The chart in Fig. 46.6 demon-

Fig. 46.7 The clue of a
hidden planet absent Bode's
law

PLANET	DISTANCE	RATIO
Mercury	3.9	===
Venus	7.2	1.9
Earth	10.0	1.4
Mars	15.2	1.5
Jupiter	52.0	(3.4)
Saturn	95.8	1.8
Uranus	192.2	2.0
Neptune	301.1	1.6
Pluto	395.4	1.3

strates the great discrepancies between theoretical and actual values in Bode's law after the discoveries of Neptune and Pluto—given the two assumptions above.

Clearly, Bode's law is just a <u>mathematical coincidence</u> rather than an authentic and valid <u>law of nature.</u>

Only a limited number of systems (besides the Solar System) are available with which Bode's law can be tested presently. The planets Jupiter and Uranus, with their sufficient number of large moons, have probably been formed in a process similar to that which formed the planets themselves. The four big satellites of Jupiter and the biggest inner satellite, Amalthea, cling to a regular spacing with orbits that are each roughly twice that of the next inner satellite, and overall they disobey Bode's law. The big moons of Uranus have similar spacing between them and they also disobey Bode's law. Of the recent discoveries of extrasolar planetary systems (a.k.a. exoplanets), only very few have enough known planets to test whether they obey or disobey Bode's law.

But how could the tremendous success of Bode's law in discovering Ceres of the Asteroid Belt be disregarded or explained? Is there an alternative method of this discovery independently of Bode's law? Yes! Indeed there is a much simpler and quite straightforward method via the pattern in the actual spacing of the planets, measured as the set of **ratios** of the successive distances from the Sun, as seen in Fig. 46.7. Such a pattern in the **ratios** of the distances could have provided the clue that a hidden planet should exist somewhere between Mars and Jupiter, and without relying on Bode's law in any way.

Table in Fig. 46.7 depicts the ratios of distances absent any planet between Mars and Jupiter. Clearly the relatively large ratio of 3.4 for Jupiter/Mars is an outlier, possibly hinting at some anomaly in the system.

The table clearly demonstrates that Jupiter represents a quantum jump in terms distancing oneself from the Sun, hinting at a possible missing planet between Mars and Jupiter. Let us then place an imaginary planet between Mars and Jupiter using the estimated average value for the empirical set of ratios.

Fig. 46.8 Ratios appear
orderly with an added planet

PLANET	DISTANCE	RATIO
Mercury	3.9	===
Venus	7.2	1.9
Earth	10.0	1.4
Mars	15.2	1.5
Hidden P.	28.3	1.9
Jupiter	52.0	1.8
Saturn	95.8	1.8
Uranus	192.2	2.0
Neptune	301.1	1.6
Pluto	395.4	1.3

The average of the following set of seven ratios is calculated as:

$(1.9 + 1.4 + 1.5 + 1.8 + 2.0 + 1.6 + 1.3)/7 = \mathbf{1.64}$, excluding the abnormal ratio of 3.4 for Jupiter/Mars from the calculation (for the simple reason that these two planets are not believed to be adjacent, and that a hidden planet exists in between them).

Hence the estimated location of a hidden planet between Mars and Jupiter is then calculated either forward from Mars or backward from Jupiter, using the ratio 1.64.

Forward from the position of Mars as $(15.2)*(1.64) = \mathbf{25.0}$.

Backward from the position of Jupiter as in the solution to the equation: (Hidden)*(1.64) = (52.0), leading to the value of (Hidden) = (52.0)/(1.64) = $\mathbf{31.7}$.

A compromise between the value of 25.0 and the value of 31.7 is given by their average value which is $(25.0 + 31.7)/2 = \mathbf{28.3}$.

Indeed, this value is very close to the actual 27.7 value of Ceres!

For all its worth, this value is also very close to the prediction of 28.0 for Ceres given by the now defunct Bode's law.

Table in Fig. 46.8 depicts the ratios of distances when a supposedly hidden planet between Mars and Jupiter orbiting at a distance of 28.3 is included in the list, showing a smooth and orderly set of ratios without any outliers or abnormal values. The more precise ratios of Hidden/Mars and Jupiter/Hidden are 1.857 and 1.839 respectively, but are rounded as 1.9 and 1.8 in table.

In conclusion: The supposed discovery of a hidden planet between Mars and Jupiter could have been made independently of the formulation of Bode's law!

Note There is an alternative way of obtaining almost the same 28.3 value for that hidden planet, namely by considering in isolation the progression of the distances from Mars to Jupiter, which is 15.2 to 52.0, and which points to the associated 3.4 ratio. The square root of 3.4 could be considered as the "midway ratio" from Mars to that hidden planet, as well as from that hidden planet to Jupiter, so that the ratio

$\sqrt{3.4} = 1.848$ is applied twice, cumulatively yielding $1.848 * 1.848 = \sqrt{3.4} * \sqrt{3.4}$ $= 3.4$. From the position of Mars, this method yields the distance of $(15.2)*(1.848)$ $= \mathbf{28.2}$ for the position of that hidden planet.

From the position of that hidden planet, this method also yields the distance of $(28.2)*(1.848) = 52.0$ for the position of Jupiter.

Let us now compare Bode's law with Kepler's 3rd law:
Why did Kepler's 3rd law turn out to be so successful, while Bodes's law failed so miserably? In order to answer this question let us look into some profound differences between the two laws:
Firstly, Bode's law is vague and fuzzy regarding what types of objects in the Solar System are to be considered. Indeed it is totally silent about the issue. Surely one cannot include and assign **n**-rank to all the little rocks and small asteroids out there. Exact and rigorous definition of what constitutes a planet is needed here in order for the law to be empirically tested. In contrast, Kepler's 3rd law does not depend on any ranking system whatsoever, as it applies to any types of objects in the Solar System, including proper planets, smaller planet—like objects, asteroids, big rocks, and so forth.
More importantly, Bode's law uses exact deterministic-type equation to describe random phenomenon, namely the random process of forming planets and asteroids from the solar nebula, the disc-shaped cloud of gas and dust left over from the Sun's formation. This is wrong!
Kepler's 3rd law uses exact deterministic-type equation to describe an exact deterministic phenomenon involving inertia and gravitational force. This is correct!
Bode's law refers to the random process of particles, gas, and dust circling a star, and eventually forming planets. Rarely would two planets form in a very close proximity to each other. Random gravitational forces ensure that things that are very near each other would eventually merge together into a single planet or a single object. This establishes some (random) distance between planets, and empirically this distance (ratio of outer planet to the next inner planet) runs roughly from 1.3 to 2.0.
Let us replace Bode's law with a more realistic model.
What is necessary here is to take a statistical and random approach to positions of planets around their stars. It is necessary to collect data on thousands or millions of star/planets systems throughout the galaxy and record relative orbital radii; then at a minimum calculate the average or the median of the ranked set of ratios of radii; or perform more some complex and appropriate statistical procedures. This is not feasible yet in our current epoch with our relatively limited technology which can only peek fuzzily at the details of other star systems, although this might become possible in the distant future.
The fallacy of the eighteenth century formulation of Bode's law is that it took only one single sample (the Solar System) instead of taking thousands or millions of samples. Bode then took this singular set of radii (one sample) all too seriously, giving it wholly undeserved importance, excessively analyzing it, trying to fit it into

a formula, instead of waiting to collect many more samples (star systems). Indeed, the right approach is to let go altogether of any exact deterministic formulation, even if an excellently fitting one is found, and view planet formations as statistical and random processes, deserving of histograms, densities and confidence intervals instead.

Bode passed away in 1826, in peace with himself, and assured of the correctness of his discovery, well before Bode's law fell apart later with the discoveries of Neptune in 1846 and of Pluto in 1930. Bode surely did not turn in his grave much during these upsetting planetary discoveries as he could certainly find plenty of comfort and solace knowing that he share the same fate and misfortune with his German compatriot— the younger Kepler (before his work on the three laws of planetary motion) who also attempted in vain to find 'explanations' for the set of distances of the planets via his two failed geometrical constructions of inscribed and circumscribed circles around regular polygons, and inscribed and circumscribed spheres around Platonic solids. Contrary to all these failed attempts by Bode and Kepler, the success of Kepler's third law can also be explained in terms of it being a much more profound quest of finding regularities and patterns in relating the two different variables of distance and time to each other in a way that would hold true across all planets, instead of finding illusive explanations for the (mostly insignificant) numerical set of values for the single variable of the distances from the Sun.

Surely, some misguided scientists could attempt to rescue Bode's approach and search for a superior deterministic equation that would embrace Neptune and Pluto as well; and surely this could be done 'successfully' if one is sufficiently inventive and looks hard enough for a supposed (illusive) pattern here, but this is not the right approach.

For example, one could apply the generic and more applicable formula using values other than 4, 3, 2 for parameters a, b, c, and with a totally different monotonic sequence that might seem more suitable:

$$\text{Distance} = a + b * c^{\{\text{Monotonic Sequence}\}}$$

This more generic formula could in principle encompass much broader physical phenomena, rightly or wrongly. A thorough search into what three parameters are needed and which monotonic sequence is best here in order to include Neptune and Pluto in the set, would almost surely yield an acceptable fit by that errant and overly enthusiastic scientist, who would soon be totally dismayed and disappointed when told of the existence of Kuiper belt and the object Eris in the Solar System, and which would almost certainly ruin the newly found good fitness. And that scientist, if he or she is sufficiently persistent, might continue the investigative process by modifying again the parameters and the monotonic sequence in order to incorporate these two new objects, and would probably find another 'successful' formula. But all this is quite futile and meaningless endeavor, because the process is random in nature and not suitable for any deterministic formulation using only the single 'sample' of our Solar System.

It should be noted that in addition to the generic formula above involving parameters a, b, c, there exist in principle infinitely many other expressions and formula types which could potentially cover (fit to data) much broader physical phenomena.

There is some exceedingly valuable and generic lesson to learn from the saga of the rise and fall of Bode's law, a lesson which should reverberate through all (deterministic and random) scientific disciplines whenever the researcher attempts to formulate whatever formula, law, rule, or theory, from a set of physical and experimental data.

Chapter 47
Developments in the Sciences Following Newton's Discoveries

Significant progress, discoveries, and major breakthroughs, in chemistry, biology, geology, sociology, psychology, economics, molecular biology, medicine, and other scientific disciplines, all came after Newton; after physics was discovered; after the scientific method was adopted; after mathematics was endowed a central role in science by Kepler, Galileo, and Newton; and after Descartes emphasized the mechanical and causality aspects of nature. It is certainly reasonable to state that the decisive success of physics in the 17th century gave inspiration and confidence to later pioneers in the sciences; the chemists, the biologists, the geologists, the psychologist, the economists, and the others, that they should indeed investigate and theorize, that there was surely something essential out there to be discovered, and that the hidden secrets of nature can be uncovered, just as it occurred in physics. It is in that sense that the work of Kepler, Galileo, and Newton, as well as Copernicus, Tycho Brahe, and Descartes, consequently and practically directly, promoted, inspired, and led to the progress of scientific discovery in general, well beyond physics.

Major developments in chemistry followed almost immediately after Newton's midwifing the birth of physics. Robert Boyle (1627–1691) was an English scientist largely regarded as the first modern chemist, best known for Boyle's law regarding the relationship between pressure, volume, and temperature of gas. Henry Cavendish (1731–1810) was an English scientist noted for his discovery of the hydrogen. Joseph Priestley (1733–1804) was an English scientist who discovered oxygen, and he is considered as one of the founding fathers of chemistry. Antoine Lavoisier (1743–1794) was a French chemist who discovered the principle of the conservation of mass, and the role oxygen plays in combustion, respiration, and rusting. John Dalton (1766–1844) was an English chemist best known for articulating and introducing the atomic theory in chemistry. Amedeo Avogadro (1776–1856) was an Italian scientist who discovered that atoms of the elements combine to form molecules. Dmitri Mendeleev (1834–1907) was a renowned and deeply admired Russian chemist best known for formulating the Periodic Law of the Elements, leading to what is known nowadays as The Periodic Table.

© The Editor(s) (if applicable) and The Author(s), under exclusive license to Springer Nature Switzerland AG 2020
A. E. Kossovsky, *The Birth of Science*, Springer Praxis Books,
https://doi.org/10.1007/978-3-030-51744-1_47

As much as astronomy needed a telescope, biology needed a microscope. The innovative use of the telescope by Galileo, and later by Kepler and others, was a major step in advancing our knowledge of the big universe within which we reside. In the same vein, the innovative use of the microscope was a major step in advancing our knowledge about the tiny cells and other microscopic entities of which we are all made of. The earliest microscopes were developed in the Netherlands in the late 16th century, and later were put to use in biological studies from around the middle of the 17th century. Zacharias Janssen (1580–1638) was a Dutch lens-maker who invented the first compound microscope in 1595. Antonie Van Leeuwenhoek (1632–1723) was a Dutch scientist who greatly improved the design of the microscope in 1664. Leeuwenhoek observed protozoa in 1676, spermatozoa in 1677, and bacteria in 1683. Matthias Schleiden (1804–1881) was a German botanist and co-founder of the cell theory, who concluded that all plants are made of cells. Theodor Schwann (1810–1882) was a German physician and physiologist who extended the cell theory to animals, and postulated that cells are the elementary particles of life. Rudolf Virchow (1831–1902) was a German biologist, known for stating that cells can arise only from pre-existing cells. Gregor Mendel (1822–1884) was a Czech-German scientist known as the founder of the modern science of genetics, who postulated the laws of Mendelian inheritance. Charles Darwin (1809–1882) was an English naturalist and biologist, best known for his contributions to the science of natural selection and his theory of evolution. Louis Pasteur (1822–1895) was a French biologist, microbiologist and chemist renowned for his discoveries of microbial fermentation and pasteurization. Pasteur provided direct support for the germ theory of disease. The discovery of the double helix, the twisted-ladder structure of deoxyribonucleic acid or DNA in 1953 by James Watson (1928–present), Francis Crick (1916–2004), as well as Rosalind Elsie Franklin (1920–1958) who also contributed to the study, was a milestone in the history of biological science. The discovery gave rise to modern molecular biology, an essential and immensely applicable body of knowledge concerning with how genes control the chemical processes within the cells.

Post-Newtonian development in physics accelerated, leading to success after success in theoretical discoveries as well as in practical applications. Newton's theory was consistently confirmed in countless observations without fail, and its theoretical body of knowledge widened and grew exponentially. In the latter part of the 18th century, the study electromagnetism began, followed by many groundbreaking discoveries. In addition, around the middle of the 19th century, major discoveries in the field of thermodynamics occurred. Both fields—electromagnetism as well as thermodynamics—yielded tremendous practical and industrial applications.

Henry Cavendish of England and Charles-Augustin de Coulomb of France around 1770–1790 established the foundations of electrostatics, regarding the repulsive force of like electric charges, and the attractive force of opposite charges.

Hans Christian Ørsted of Denmark demonstrated in 1820 that electric currents create magnetic effects, thus discovering the first connection between electricity and magnetism. Current can be thought of as moving electrons, namely as moving negative charges, and that motion changes the static electric field around them, which in turn creates magnetic effects.

André-Marie Ampère of France in 1820 identified electric currents as the source of all magnetism, including permanent metal magnets. Ampère showed that two parallel wires carrying electric currents attract each other as magnets if currents flow in the same direction, and repel each other as magnets if currents flow in the opposite directions. In summary, what Ampère demonstrated is that it is only the movement of electric charges (i.e. current) that can induce magnetic effects.

Michael Faraday of England in 1831 discovers that changing (time varying) magnetic fields are also sources for electric effects. This is called Electrical Induction. His statement implies that by moving a metal magnet next to a wire, electrical current can be induced within the wire, because the changing magnetic field causes an electric effect which moves the electrons in the wire and thus inducing current.

It was James Clerk Maxwell of Scotland who put it all together, first by concisely and mathematically summarizing all that was discovered in the field before him, including his own additions and interpretations. This led to what is today famously known as the four 'Maxwell's Equations'. In 1864 Maxwell studied how electric and magnetic fields travel through space as waves, reasoning that a rapid acceleration of a physical metal magnet would induce a change in the electric field ahead of it, and which in turn would induce a further change in the magnetic field ahead of it, and so forth. To his great astonishment, the speed of that wave was mathematically calculated (almost exactly) as the speed of light! This was one of the most dramatic moments in the whole history of science, comparable to the dramatic moment of Newton's discoveries under the broad apple tree. Light was suddenly understood then as an electromagnetic phenomenon! Maxwell's mathematical discovery regarding the true nature of light has contributed enormously to humanity's economical, social, and scientific development, since our entire mode of communication via telegraph, radio, television, or the microwave, is based wholly on his discovery. Maxwell's work was nothing but theoretical scriptions of equations and the manipulations of abstract formulas on papers with a pencil. It was Heinrich Rudolf Hertz (1857–1894) of Germany who in 1887 attempted to conclusively prove Maxwell's light theory by demonstrating the existence of the electromagnetic wave in a physical setting. Hertz induced and sparked electricity in one wire, only to observe the electromagnetic wave manifesting itself in another wire nearby. This was the first ever (primitive) radio 'communication'. Hertz also managed to confirm that the phenomenon shared several properties with light such as reflection and interference, as predicted by Maxwell.

Interestingly, Maxwell's Equations do not obey the principle of Galilean Relativity, and this fact appeared puzzling and contrary to the physics branch of mechanics which do obey this relativity principle. This is why Maxwell's Equations greatly inspired Albert Einstein in developing his theory of Special Relativity. This dichotomy between the two major fields in physics prompted Einstein to re-define the basic concepts of space, time, speed, and simultaneity, based on a more solid foundation, and which then allowed for the unification of all physical phenomena, mechanics as well as electromagnetism, so that both would obey the relativity principle equally.

Let us now turn our attention to thermodynamics, which is a branch of physics concerned with the energy associated with heat, heat transfer, temperature, and the relation to radiation and properties of matter. Sadi Carnot 1824 publication of his immensely perceptive memoir titled 'Reflections on the Motive Power of Fire' initiated the study of thermodynamic as a modern science. James Joule in 1843 experimentally discovered the mechanical equivalent of heat and this experiment demonstrated the validity of the first law of thermodynamics, namely that energy can neither be created nor destroyed, but can only be transferred or changed from one energy form to another form, such as heat, mechanical work, or radiation. Building on James Joule's work regarding the conservation of energy, Rudolf Clausius formulated the second law of thermodynamics in 1850, namely that heat cannot spontaneously flow from a colder location to a hotter location. In 1865 Clausius gave the first mathematical version of the concept of entropy, stating that the entire energy of the universe is fixed, while the entire entropy of the universe generally increases and tends to a maximum. In addition, beginning in 1859, James Clerk Maxwell investigated the kinetic theory of gases, viewing gas simply as a collection of fast moving atoms and molecules, each with its distinct speed and direction, and all constantly colliding with one another randomly. Expanding on Maxwell's work, Ludwig Boltzmann further related the kinetic theory to the second law of thermodynamics leading to the Maxwell–Boltzmann distribution. Statistical mechanics attempts to explain and predict the thermodynamics macro properties of solids, liquids, and gases, in terms of the micro statistical configurations of its constituent molecules, pertaining to the average molecular speed and rotation, as well as the molecular mass. The core insight of statistical mechanics applied to thermodynamics is that heat and temperature simply reflect the average speeds of the moving, rotating, or vibrating molecules constituting the substance; that in essence heat and temperature ultimately measure motion. Even though most chemists in the 19th century shared Boltzmann's and Maxwell's belief in the actual existence of atoms and molecules ever since Dalton articulated the atomic theory in chemistry, physicists generally were still very skeptical of the notion until Einstein's 1905 publication on Brownian-Motion, followed by experimental verification of his paper, thus establishing the existence of atoms and molecules as a scientific fact.

At the dawn of the 20th century, Albert Einstein singlehandedly developed Special Relatively in 1905 which reduced Newton's laws of motion to the special case of a world which moves much slower than the speed of light. This special case includes all of our daily experiences with motion as well as the motions of the planets, the Moon, the comets, and so forth, all which do not truly move fast when compared with the speed of light. Einstein's purely theoretical and elegant work in attempting to unify all branches of physics led to the famous mass-energy conversion expression $E = mc^2$. In 1915 Einstein developed General Relatively which profoundly altered and upended Newton's description of universal gravitation.

Substantial studies and experiments in Quantum Mechanics started around 1900 with Max Planck (1858–1947) in Germany, and Ernest Rutherford (1871–1937) in England. Niels Bohr (1885–1962) from Denmark then made foundational contributions to the understanding of atomic structure and quantum theory, followed by James Chadwick (1891–1974) of England, and Enrico Fermi (1901–1954) of Italy who made significant contributions to the development of statistical mechanics, quantum theory, and nuclear physics. A series of experiments on nuclear fission by Lise Meitner, Otto Hahn, and Fritz Strassmann in late 1938 and early 1939 in Germany, where uranium was bombarded with neutrons, had managed to split it into two smaller atoms, accompanied by the ejection of rapidly moving neutrons and a large amount of energy release.

Lise Meitner's insights were key to the theoretical understanding and the wider implications of these experiments, as she demonstrated that these results constituted the first ever empirical confirmation of Einstein's theoretical formula $E = mc^2$. This was accomplished by measuring and explaining the source of the tremendous release of energy in atomic decay. These experiments showed that resultant combined masses after the split were lighter than the original masses, and that the discrepancy or loss in the mass was due to the conversion of that mass into energy, corresponding nearly exactly to the $E = mc^2$ relationship. Albert Einstein was delighted with this experimental confirmation of his theoretical work, and fondly praised Meitner as 'Our German Marie Curie'.

Meitner, Hahn, and Strassmann, published the results of their experiments in January and February of 1939, and all this had electrifying effects on the international scientific community, because these results indicated that there was now a possibility that fission could be used as a weapon of war, and that the knowledge was then mostly in German hands.

Two very distinct and totally unrelated historical events converged at about the same time around 1939, and which unfortunately conspired against humanity and led to the creation of the atomic and thermonuclear bombs. The first event was the emerging threat of resurgent and emboldened Nazi Germany in Europe, as well as the fear of the constantly expanding Empire of Japan in the East, eventually leading to the Second World War. The second event was the very rapid progress just around the same time in Quantum Mechanics, and the realization of the enormous energy release associated with the splitting of heavy atoms. The convergence of these two events was a truly remarkable coincidence, as it was historically timed almost precisely so as to produce the greatest adverse effect!

The Hungarian physicist Leo Szilard conceived of the possibility of nuclear chain reaction, and patented the idea of a nuclear fission reactor in 1934. Szilard realized later in 1939 that the fission of heavy atoms could potentially be used for military purposes, and that the Nazi regime may become involved. In August 1939, Leo Szilard and fellow Hungarian refugee physicists Edward Teller and Eugene Wigner persuaded fellow refugee Albert Einstein who fled Nazi Germany and moved to the United States to lend his name to a letter directed to President Franklin Roosevelt urging him to have the US build such a bomb first before Germany might acquire it. President Roosevelt obliged, allocating money and resources for a project to create such a bomb. This led to the formation of the Manhattan Project during the years 1942–1946, and the creation of the first atomic bombs which were used against Japan in August 1945.

Had progress in Quantum Mechanics arrived much sooner, say well before World Wars I and II, or had its progress been slower, arriving after the wars, no great motive would exist for governments to go to such extreme lengths and initiate such huge, costly, and uncertain project, and most scientists would probably never accept using their ideas and expertise to build such powerful and destructive bombs. Most likely, no bomb would have been developed at all in such scenarios.

For example, Albert Einstein himself was a well-known pacifist, and in late 1914 around the beginning of the First World War he clashed with his colleagues in the Prussian Academy of Sciences of the University Berlin, rallying against militarism, nationalism, and especially against the war. In October 1914 the German physiologist Georg Friedrich Nicolai wrote "Manifesto to the Europeans", a poignant and magnificent anti-war pamphlet, denouncing the senseless, absurd, and highly destructive European war occurring on this cultured, civilized, and scientifically-advanced continent. It was signed by Albert Einstein and two other German scholars, but then remained unpublished for lack of more participants, until it was subsequently brought to light by Einstein years later. Clearly, Einstein would never have written a letter to the US president urging him to develop weapons of mass destruction if not for the perceived urgency caused by the possibility that Nazi Germany would develop such weapons first, given that so many German physicists were already working and excelling in the field.

Putting aside political, historical, and military concerns, as well as questions regarding the moral dilemma of building a highly destructive and horrific bomb initially intended against the highly amoral, barbaric, and extremely dangerous regimes of Nazi Germany and the Empire of Japan—this colossal scientific and engineering endeavor of the Manhattan Project was an extraordinary and unique event in the whole history of science! It gathered the best minds and talents, the most capable and innovative scientists and leading figures in the fields of mathematics, physics, chemistry, engineering, and computing, settling them all together in the small town of Los Alamos, New Mexico, southern United States, where they exchanged ideas and constantly interacted with each other, and with no worries about providing food and shelter or the need for work, with enthusiasm and idealism in the belief that their cause was just, worthy, and noble. It led to many discoveries and advances in many fields of science as well as mathematics and statistics, far beyond the original goal of building the atomic bomb. Discounting the horrific death and destruction that came later, The Manhattan Project evokes fond memories of the beautiful painting of the 'School of Athens' by the Italian Renaissance artist Raphael, or imaginary scenes of large gathering at the Library of Alexandria of scientists, mathematicians, philosophers, and poets, such as Archimedes, Ptolemy, Hypatia, Euclid, and Eratosthenes, in a city regarded at that era as the capital of knowledge and learning. What new discoveries and insight could have been obtained by just several such gatherings and free discourse of the greatest minds of ancient Greece!

The scientific revolution and the great advancement in technology had made enormous contribution to the quality of our life. We benefit from cars, telephones, airplanes, medicine, enormous increase in food production, and essential life-saving surgeries performed under complete anesthesia without the slightest of pain. Our life expectancy had doubled approximately since the time Newton passed away to the present era, increasing from around 40 years to around 80 years. Longevity of sorts had been achieved.

Yet, as humanity vastly advanced in the fields of physics, chemistry, and engineering, weapon production ability had been increasing step-by-step accordingly, culminating in the Manhattan Project and its successful splitting of the uranium and the plutonium atoms, followed by Hiroshima and Nagasaki. Shortly afterwards in 1951–2 came the building of thermonuclear weapons, and the fusing of the hydrogen atoms, boldly imitating the mighty stars right here on little planet Earth. Consequently, we have arrived at the realization that an enormous danger regarding the possibility of total annihilation is hanging over humanity. One then naturally wonders whether or not in the final analysis, humanity has truly benefited from science, technology, and knowledge. Are we truly better off today being much more knowledgeable and capable than in our ancient eras of ignorance and simplicity when the fate of humanity was assured, barring the possibility of some rare natural calamities, such as geological catastrophes, super volcanic eruption, or the striking of a massive meteor from outer space onto the Earth's surface.

Would Kepler, Galileo, and Newton approve of humanity's present condition? Would they lament and regret the bad use of their intellectual work and achievements for building such massively destructive tools capable of annihilating humanity itself? The author dares to hope against all hope that the birth of science would never engender the death of humanity.

References

Bode JE (1768) Anleitung zur Kenntniss des gestirnten Himmels (Instruction for the knowledge of the Starry Heavens)

Brown K (2017) Reflections on relativity. http://www.mathpages.com/rr/s8-01/8-01.htm lulu.com, 4 Nov 2017. ISBN 10:1257033026

Caspar M (1993) Kepler. Dover Books on Astronomy, Paperback, 8 Sept 1993. Until his death in 1956, Professor Max Caspar was the world's foremost Kepler scholar

Copernicus N (1543) De revolutionibus orbium coelestium (On the revolutions of the heavenly spheres), Original Nuremberg edn. Johannes Petreius, Nuremberg

Copernicus N (1976) On the revolutions of the heavenly spheres. Translated by A.M. Duncan. David & Charles. ISBN 0-7153-6927-X.

Copernicus N (1995) On the revolutions of heavenly spheres. Great Minds Series, translated by Charles G. Wallis. ISBN 10: 1-57392-035-5

Descartes R (1627) Regulae ad directionem ingenii (Rules for the direction of the mind). Incomplete. First published posthumously in Dutch translation in 1684 and in the original Latin at Amsterdam in 1701 (R. Des-Cartes Opuscula Posthuma Physica et Mathematica). The best critical edition, which includes the Dutch translation of 1684, is edited by Giovanni Crapulli. Martinus Nijhoff, The Hague, 1966

Descartes R (1637a) Discours de la method (Discourse on the method). An introduction to the Essais, which include the Dioptrique, the Météores and the Géométrie

Descartes R (1637b) La Géométrie (Geometry). Descartes' major work in mathematics. An English translation was done by Mahoney M (1979) Dover, New York

Descartes R (1641) Meditationes de prima philosophia (Meditations on first philosophy). Also known as Metaphysical Meditations

Descartes R (1644) Principia philosophiae (Principles of philosophy). A Latin textbook at first intended by Descartes to replace the Aristotelian textbooks then used in universities.

Donahue HW (1993) Johannes Kepler new astronomy. Translation of Kepler 1609 work. Cambridge University Press

Drake S (1973) Galileo gleanings XXII: Galileo's experimental confirmation of horizontal Inertia, vol 64, pp 291–305. Unpublished Manuscripts. Isis

Drake S (1957) Discoveries and opinions of Galileo. Doubleday & Company.

Drake S (Author), Levere HT , Swerdlow MN (eds) (2000) Essays on Galileo and the history and philosophy of science, vol III. University of Toronto Press, Scholarly Publishing Division

Galileo G (1632) Dialogue concerning the two chief world systems: ptolemaic and copernican. Translated by Drake S, Gould SJ et al (2001, October 2) Modern library, New edn.

© The Editor(s) (if applicable) and The Author(s), under exclusive license to Springer   215
Nature Switzerland AG 2020
A. E. Kossovsky, *The Birth of Science*, Springer Praxis Books,
https://doi.org/10.1007/978-3-030-51744-1

Galileo G (1638) Discorsi e dimostrazioni matematiche, intorno à due nuove scienze (Mathematical discourses and demonstrations, relating to two new sciences). Translated by Crew H, de Salvio A (1914) Re-published in 1954. Dover Publications Inc., New York. ISBN 978-0-486-60099-4. Now available online via the link: http://galileoandeinstein.physics.virginia.edu/tns_draft/index.html

Galileo G (1638) Two new sciences/a history of free fall. Translated in 2000 by Drake S Wall & Emerson, Inc. ISBN 13:9780921332503.

Gingrich O (1975) The origins of Kepler's Third Law. Beer A, Beer P (eds) Vistas in astronomy, vol 18, pp 595–601. Link at https://www.sciencedirect.com/https://www.sciencedirect.com/

Giusti E (1998) Galileo's De motu antiquiora (The older writings on motion). Nuncius 13(2):427–60. Galileo Galilei's early written work on motion. It was written largely between 1589 and 1592, but was not published until 1687, after his death

Kepler J (1596) Mysterium cosmographicum (The cosmographic mystery)

Kepler J (1981) Mysterium cosmographicum (The secret of the universe). Translated by Duncan AM Abaris books, New York

Kepler J (1604) Astronomiae Pars Optica (The optical part of astronomy)

Kepler J (1609) Astronomia Nova (New astronomy). Translated by Donahue WH (1992) Cambridge University, Cambridge. ISBN 0-521-30131-9

Kepler J (1619) Harmonices Mundi (The harmony of the world)

Kepler J (2011) Harmonies of the world, Kindle edn, translated by Wallis CG. ISBN 1496085167

Koestler A (1990) The sleepwalkers: a history of man's changing vision of the universe (compass). Penguin Books

Newton I (1669) De analysi per aequationes numero terminorum infinitas (On analysis by equations with an infinite number of terms). Written in 1669, published in 1711

Newton I (1684) De motu corporum in gyrum (On the motion of bodies in an orbit)

Newton I (1687) Philosophiae Naturalis Principia Mathematica (Mathematical principles of natural philosophy)

Newton I (1704) Opticks: a treatise of the reflexions, refractions, inflexions and colours of light

Newton I (1736) Method of fluxions. The book was completed in 1671 and published posthumously in 1736. Fluxion is Newton's term for a derivative

Newton I (1754) An historical account of two notable corruptions of scripture. This was sent in a letter to John Locke on 14 November 1690 and built upon the textual work of Richard Simon and his own research. The text was first published in English in 1754, 27 years after Newton's death

Owen G (2005) The book nobody read: chasing the revolutions of Nicolaus Copernicus. Penguin Books

Owen G (2006) God's universe, 1st edn

Palmieri P (2007) Galileo's experiments with pendulums: then and now. Phil Sci Archive. Available at http://philsci-archive.pitt.edu/3549/

Index

Printed in the United States
By Bookmasters